Ornamental Plants for Gardening

V.L. Chopra
Markandey Singh
CSIR-Institute of Himalayan Bioresource Technology,
Palampur (H.P.), India

SCIENTIFIC
PUBLISHERS (INDIA)

Published by:

Scientific Publishers (India)
5 A, New Pali Road, P.O. Box 91
Jodhpur 342 001 (India)

Branch Office
Scientific Publishers (India)
4806/24, Ansari Road, Daryaganj
New Delhi - 110 002 (India)

E-mail: info@scientificpub.com
Website: www.scientificpub.com
Print, 2021

ISBN: 978-81-7233-830-5

Typeset by Rajesh Ojha
Printed in India

PREFACE

Intricacies of nature, expressed through vegetative and floral diversity, have continued to fascinate human curiosity through the ages. The intense love for nature of the tribal communities, who make special efforts to integrate items of this diversity in their living quarters, is a telling example of this instinct. The expectations for gardening requirements of the urban gardeners are more demanding and sophisticated. This is particularly so for those living in towns, away from pristine environments. This group desperately seeks information and advice on what to grow and how to properly maintain what is grown. This information is not readily available specialy for those wishing to create an attractive and novel garden in the hilly regions. The book "Ornamental Plants for Gardening" is an attempt to fill this gap. In fourteen chapters, the book covers all aspects of gardening such as principals of garden designing; details of different groups of ornamentals; their varieties and cultivation practices that will optimally express the aesthetic potential of what is chosen for planting; diseases and pests and how to control them; practices for production and storage of seeds/planting material; tools and equipment required for effective gardening and where to source them from. Special attention has been given to information needs of home gardeners of the western Himalayan hill areas. Each chapter has been prepared by a recognized expert in the relevant subject to ensure quality and authenticity of information.

It is a pleasure to record gratitude to many who have helped in developing and executing this project. Authors who have been so generous with their expertise and time top this list. Dr. Paramvir Singh Ahuja was generous and helpful in numerous ways. As conduit for dealing with the publishers, Shri Mukhtiar Singh provided a very smooth channel and we thank him for this valuable input. At a personal level, VLC is very grateful to Dr. SK Vats and Shri JK Prashar but for whom it would have been impossible for him to be effective. Finally, we would like to thank the publishers for their understanding and cooperation.

V.L. Chopra
Markandey Singh

Palampur
October ----, 2012

CONTRIBUTORS

Dr. B. K. Banerji
Sr. Principal Scientist & Head
Floriculture Division
CSIR- National Botanical Research Institute (NBRI),
Rana Pratap Marg,
Lucknow-226001 (U.P.)

Dr. D. Mukherjee
Former Head
Floriculture Division
CSIR-IHBT, Palampur (H.P.)
BK-6 Sector- 2 Salt Lake
Kolkata-1733091 (W.B.)

Dr. Desh Raj
Associate Professor
Department of Horticulture
CSKHPAU,
Palampur- 176061 (H.P.)

Mr. Devendra Dhyani
Sr. Principal Scientist
Floriculture Division
CSIR- Institute of Himalayan Bioresource Technology (IHBT),
Palampur-176061 (H.P.)

Dr. Y.C. Gupta
Professor & Head
Department of Floriculture and Landscaping, Dr. Y.S. Parmar University of Horticulture and Forestry, Nauni,
Solan-173230 (H.P.)

Dr. F. U. Khan
Associate Professor
Sher-e- Kashmir University of Agricultural Sciences and Technology of Kashmir,
Shalimar,
Srinagar-191121 (J&K)

Dr. I. T. Nazki
Assistant Professor
Sher-e- Kashmir University of Agricultural Sciences and Technology of Kashmir, Shalimar,
Srinagar-191121 (J&K)

Dr. Kanwar Pal Singh
Principal Scientist
Division of Floriculture and Landscaping, Indian Agricultural Research Institute, Pusa,
New Delhi-110012

Dr. Markandey Singh
Principal Scientist
Floriculture Division
CSIR- Institute of Himalayan Bioresource Technology (IHBT),
Palampur-176061 (H.P.)

Dr. Mast Ram Dhiman
Sr. Scientist
Indian Agricultural Research Institute, Pusa, Regional Station
Katrain
Kullu-175129 (H.P.)

Mr. Bhavya Bhargava
Ph D Scholar
Department of Floriculture and Landscaping, Dr. Y.S. Parmar University of Horticulture and Forestry, Nauni,
Solan-173230 (H.P.)

Dr. Namita
Scientist
Division of Floriculture and Landscaping, Indian Agricultural Research Institute, Pusa,
New Delhi-110012

Dr. Neelofar
Associate Professor
Sher-e- Kashmir University of Agricultural Sciences and Technology of Kashmir, Shalimar, Srinagar-191121 (J&K)

Mr. P. Pavan Kumar
Ph D Scholar
Division of Floriculture and Landscaping, Indian Agricultural Research Institute, Pusa, New Delhi-110012

Priyanka Sharma
Ph D Scholar
Department of Floriculture and Landscaping , Dr. Y.S. Parmar University of Horticulture and Forestry, Nauni, Solan-173230 (H.P.)

Dr. R.K. Roy
Principal Scientist
Botanic Garden
CSIR-National Botanical Research Institute (NBRI), Rana Pratap Marg, Lucknow-226001 (U.P.)

Dr. S.P.S. Raghava
Former Project Coordinator (Floriculture), AICFIP, ICAR, Present Address: 3/80, Sector -2, Rajender Nagar, Sahibabad, Ghaziabad -201005 (U.P.)

Dr. S.S. Sindhu
Principal Scientist
Division of Floriculture and Landscaping, Indian Agricultural Research Institute, Pusa, New Delhi-110012

Mr. Sanjay Kumar
Technical Assistant
Floriculture Division
CSIR- Institute of Himalayan Bioresource Technology (IHBT), Palampur-176061 (H.P.)

Dr. Satish Kumar
Technical Assistant
Botanic Garden
CSIR-National Botanical Research Institute (NBRI), Rana Pratap Marg, Lucknow-226001 (U.P.)

Dr. T. Janakiram
Principal Scientist & Head
Division of Floriculture and Landscaping, Indian Agricultural Research Institute, Pusa, New Delhi-110012

Dr. V.S. Raju Dantuluri
Sr. Scientist
Division of Floriculture and Landscaping, Indian Agricultural Research Institute, Pusa, New Delhi-110012

Dr. Z. A. Bhat
Assistant Professor
Sher-e- Kashmir University of Agricultural Sciences and Technology of Kashmir, Shalimar, Srinagar-191121 (J&K)

Dr. Z. A. Rather
Assistant Professor
Sher-e- Kashmir University of Agricultural Sciences and Technology of Kashmir, Shalimar, Srinagar-191121 (J&K)

CONTENTS

CLASSIFICATION AND PRINCIPLES OF HOME GARDENS

R. K. Roy and Satish Kumar

INTRODUCTION AND IMPORTANCE

Home is much more than just a roof over head; it occupies a unique place in our social life. A lovely home garden is a fascinating adjunct. It is a source of revitalizing both body and mind. Whenever a new house is planned, an attempt is made to provide space for a patch of lawn and some plants to make the house lively.

Our association with plants goes back to time immemorial and has driven us to grow plants in and around the house. Therefore, home gardening refers to growing plants in a systematic manner in and around the house. These plants include ornamental plants for aesthetics and plants such as vegetables, fruits and medicinal herbs for utility. The purpose of home gardening is basically to meet the requirements of flowers, vegetables and fruits for home consumption. The success of the garden depends on the wholehearted and active participation of family members. The joy and enjoyment associated with growing plants and to see them bearing flowers and fruits is simply beyond description.

Home gardening is easy in rural and peri-urban areas as ample space is available around the home. One can have a full-fledged garden in these areas so that the requirement of flowers, fruits and vegetables can be largely met. However, in urban areas, the scope of home gardening is limited, primarily due to paucity of available land. The system of multi-storied buildings and living in flats has virtually ruined the gardening habit of the people. The situation is very alarming in metropolitan cities where land value is very high and people cannot afford to have open space for outdoor garden. In such cases, people may go for growing plants in pots and keeping them in verandahs and on roof-tops. But the pleasure and advantage of having some ground space for gardening cannot be enjoyed. However, skilful planning and right selection of ornamental plants can bring some relief and enjoyment.

CLASSIFICATION OF HOME GARDEN

The size of the home garden is one of the important factors for deciding its features and choice of plants. Therefore, classifying home gardens is necessary for better understanding of the requirements and its planning. The following classification has been made exclusively on the basis of the available area:

Large-Sized Area

This category includes only large gardens for bungalows. Features such as rock garden, pool and pergola, which require comparatively large space are ideal for these gardens. Plantation of trees, shrubbery border and large lawns can be accommodated in these gardens.

Features: Lawn, rose garden, herbaceous border, shrubbery border, pergola, rock garden, children's corner, aquatic bodies, fountain etc. The features are selected as per the house owner's liking.

Plants: Trees may be accommodated along the boundary wall. 'Royal Palm' is recommended along the main passage. Other medium trees having fragrant and attractive flowers are also suitable. Usually all kinds of shrubs find place in large home gardens. Bulbous and rhizomatous plants in the form of beds in various designs add attractiveness to the gardens.

Medium-Sized Area

The features and plants selected are based on the size of the area and their suitability.

Features: Lawn, rose bed, beds of seasonal flowers, shrubs in groups, small lily pool etc. The features are selected mainly as per the availability of space.

Plants: Selected dwarf trees, all types of shrubs, annuals, bulbous plants, creepers suitable for training on porch, gate etc.

Small-Sized Area

In this category, very small space is available for development of the garden. Features and plants requiring less space are ideal.

Features: Lawn, limited number of roses, seasonals and bulbous plants in beds, small lily pools and potted plants

Plants: Selected shrubs, seasonals, creepers, bulbous plants, house plants.

Fig. 1. A view of an aesthetic home garden

PRINCIPLES OF HOME GARDENING

Like other subjects, garden designs are also based on certain principles. These help in proper designing and selection of right garden features, plant species, placement and allocation of space, colour scheme etc. Moreover, style, elevation, harmonization, type of the garden are also taken into account while designing the gardens. The main purpose is to achieve an aesthetic effect in totality.

Keeping these in mind, the main principles are categorized into the two groups.

Fig. 2. Equidistance planting and repetition of feature

Principles for Selecting Structure

The principles basically provide guidelines about selection of structures which are integral part of the garden.

Structures

Every structure of the garden is important and expresses style of the selected design.

- Boundary or Enclosure

This is an important element of the garden which unifies the house or structure or building and the garden. Therefore, material and design of the boundary or enclosure must be in harmony with the overall design of the house.

- Entrance or Gate

This should be in conformity with the style and principles of the design adopted.

Form and Shape

Besides selection of individual plant or groups, the contour, levels, shape and form must match with the design.

View

Placement of the structures and features of the garden should be made in conformity with certain principles so that the main objects are focused and they draw attention of the visitors. This creates a balanced view.

Unity

Unity is one of the main principles of design and is expressed by harmonious placement of garden features in a proper way. Selection, placement and repetition of any particular element help to create unity in the garden.

Repetition

This has direct relation with unity. Repetition of uniform elements having similar shape and colour creates unique harmonious effect.

Simplicity

This emphasizes on adopting simple design and avoiding crowding of objects, plants, decors etc.

Balance

As per design and style of the garden, balance should be maintained with regards to colour scheme, plants (shape, texture), structures and grades so that everything is integrated appropriately.

Colour

Colour of flowers and leaves constitutes a critical factor for design of the garden. Primary colours (violet, blue, green, yellow, orange and red) as well as secondary shades should be used either in contrast or in harmony depending upon the availability. Generally, bright colours are put closer to the visitors while cool colours at a distant part of the garden.

Proportion

Every feature should be in relation to the total space available for the garden. A small home garden should not contain voluminous ponds or giant statues which create imbalance.

Rhythm

Planning should be done to ensure the continuance of flowering in the garden. Gradual gradation of height and colour of the plant species should be adopted so that the changes are noticed smoothly.

Accent

This refers to a tree or shrub with attractive and distinguishing characters such as shape, leaf texture, flower colour and rarity that is situated in the prime location of the garden in a group or as individual to break the monotony of form, colour and arrangement of other plants in the design for creating more interest and pleasure.

DESIGN AND LAYOUT OF THE HOME GARDEN

The term 'garden design' refers to scientific combination and placement of all garden features such as walks, drives, lawns, arches, pergolas, plants etc. in such a way that every feature expresses both its individual beauty and a total effect in complete harmony and judicious blend. Designing of a garden and its layout in the field is a specialized job. It requires a series of considerations including area, climate, soil, features and above all knowledge of growth and flowering of the plants. To achieve good results, a garden should be designed carefully because design is a semi-permanent activity and frequent change of the layout is not possible. The question then arises about designing the home garden for the best utilization of available space. People are often confused about what to plant and where.

The following points are important to consider while designing a garden:

- Area available for the layout of the garden and the soil type
- Availability of basic facilities like drainage, irrigation, power supply etc.
- Natural grade of the area should be retained and used accordingly.
- Simplicity of the design according to the purpose and utility giving due consideration to owner's choice and convenience
- Avoiding inclusion of too many features and overcrowding of plants
- Different types of plants such as trees, shrubs, creepers, herbs, bulbs, annuals should be included so that seasonal changes are reflected in the garden and it remains presentable round the year.
- Judicious selection of features and allocation of space
- A good colour scheme by proper selection of plants and features is the key element of designing, which makes the garden more interesting and enjoyable.

The purpose of home garden is mainly for aesthetics but since it is around the house, utility purpose cannot be ignored. Therefore, careful thought should be given to both the aspects while designing. There are two main types of designing - formal and informal.

The main characteristics of formal designing are – symmetrical designs, straight roads, sharp edges, geometrical shapes of the flower beds (square, rectangular, round etc.), artificial grades, formal and equidistant planting.

Informal designing includes asymmetrical designs, smooth curves, free and natural contours, irregular shapes of the flower beds etc.

The design to be followed should be decided well in advance considering available area, building architecture and above all, owner's choice. The whole layout of the garden, therefore, depends on the type of design selected.

It is suggested that before taking up layout of the garden, a plan should be prepared on paper so that all features are well defined. Allocation of space to different features is also important and should be decided at planning stage. At least one third area of the garden should be under lawn and the remaining area should be allocated proportionately to other features. Selection of right type of plants is an important factor. Before selecting any plant species, following parameters should be considered carefully:

Fig. 3. Formal design

- Edaphic and agro-climatic suitability
- Height, shape and spread at maturity
- Leaf texture and growth rate
- Flower colour and flowering season
- Nature – deciduous or evergreen

Fig. 4. Informal design

In a home garden, there is little scope for planting trees. Therefore, only dwarf trees can be placed in such a way that minimum area is occupied and there is no harm due to the shadow of these trees. Preferably, flower beds should be located at the periphery of the lawn by the side of drives or paths or passages. Shrubs should be planted along the fence or boundary in the form of a border or individually. Creepers should preferably be placed near the gate, portico, fence, arch, pergola and trained over these structures to exploit floral beauty. Other features such as pools, rock gardens, seats, kiosks should be designed and planned judiciously so that they are well integrated.

Layout of the garden should be taken up when the construction of building is completed so that there is no disturbance and damage to the plants. As a general rule, all physical features such as drives, roads, paths, walks, mounds, pool as well as drainage and irrigation system should be taken up at the first stage before plantation. If any artificial grading is to be created with the help of civil work, besides electrical work or fittings and garden lights, it should also be completed before plantation. Simple and neat design of the front portion of the garden facing the gate makes it more attractive. It should not be overcrowded with plants and features. Rather, some selected plants with attractive foliage and flowers should be planted in the front portion so that it becomes impressive. Other features such as rock garden, pools, children's cornershould be planned for the back portion. The ultimate purpose is to create a balanced effect in combination with different features to make the ambience more attractive and pleasing.

Some of the basic features of the home garden are - Lawns, trees, shrubs, climbers, herbaceous borders, flower beds, arches, pergolas, rose gardens or rose beds, canna beds, bulbous beds, rock gardens, pools, children's corners, walks and paths (crazy path and stepping stones), fences, garden gates, seats, kiosks, green houses, kitchen gardens etc.

MAINTENANCE OF HOME GARDEN

Home gardens are generally categorized as intensive type of gardens that require special care and maintenance round the year. Even a little negligence is very promptly reflected in the quality of display as it is viewed very closely by the visitors. Therefore, care should be taken so that all maintenance including cleaning and removal of dead and diseased leaves and plant parts are taken up in time. Seasonal effects must be reflected in the garden by proper planning and execution.

SUGGESTED READING

1. Anonymous 1973 *Practical Guide to Home Landscaping*, Sydney: Reader's Digest Service Pvt. Ltd.

2. Arora JS 1990 *Introductory - Ornamental Horticulture*, Ludhiana, Punjab, India: Kalyani Publishers.
3. Bose TK and Mukherjee D 1972 *Gardening in India*, Calcutta, West Bengal, India: Oxford and IBH Pub. Co.
4. Bose TK, Dhua RS and Maity RG 1999 *Floriculture and Landscaping*, Calcutta, West Bengal, India: Naya Prokash
5. Crockett James Underwood 1972 *Foliage House Plants*, Virginia: Time-Life Books Inc.
6. Crowe Sylvia 1958 *Garden Design*, London: Country Life Ltd.
7. Gopalaswamiengar KS 1935 *Complete Gardening In India*, Madras, India: The Huxlery Press
8. Kachroo P and Zarabi Priya 1995 *Handbook of Rock Gardening on The Hills*, Indian Council of Agricultural Research
9. Manaker H George 1987 *Interior Plantscapes (2nd Edition)*, New Jersey, USA: Prentice Hall, Inc.
10. Nambison KMP 1992 *Design Elements of Landscape Gardening*, New Delhi: Oxford and IBH Publishing Co. Pvt. Ltd.
11. Pal BP and Krishnamurthy S 1967 *Flowering Shrubs*, New Delhi: Indian Council of Agricultural Research
12. Randhawa MS 1983 *Flowering Trees*, India: National Book Trust
13. Roy RK 1999 *Ornamental Annuals*, NBRI, Lucknow, U P, India: Economic Botany Information Service
14. Roy RK 2003 *Home Garden*, NBRI, Lucknow, U P, India: Economic Botany Information Service
15. Roy RK 1988 *Lawn: A Basic Feature of Every Garden*, Flower Show Bulletin, Calcutta Flower Growers' Asso. **4** 82-85
16. Roy RK, Kumar R and Datta SK 2006 *House Plants*, NBRI, Lucknow, U P, India: Economic Botany Information Service
17. Sommers P 1982 The UNICEF Home Gardens Handbook: For People Promoting Mixed Gardening in the Humid Tropics, UNICEF
18. Sharma BS 1993 *A Guide to Home Gardening*, New Delhi: Publication Division, Min. Information and Broadcasting, Government of India
19. Trivedi Pratibha 1996 *Home Gardening*, New Delhi: Indian Council of Agricultural Research
20. Swarup Vishnu 1997 *Garden Flowers*, New Delhi: National Book Trust
21. Wright Michel 1978 *The Complete Book of Gardening*, Ebury Press and Michel Joseph Ltd.

BULBOUS ORNAMENTAL PLANTS

T. Janakiram, Namita and P. Pavan Kumar

INTRODUCTION

The enormous genetic diversity of bulbous crops offers immense potential as cut flowers, pot plants or bedding plants, landscaping and decorations. These crops are used for consumption as food and/or medicine, fibre and other material useful to mankind. International trade of these flower crops is rising rapidly. The ornamental genotypes such as bulbs, tubers, rhizomes and tuberous roots are collectively called flower bulbs (Halevy 1990, Rees 1989). These four types of storage structures accumulate food during growing seasons and enable the plant to survive for a long dormancy. Fresh foliage resumes as soon as weather conditions become conducive. A bulbous plant normally has three phases in a year - growth, flowering and dormancy. Vegetative growth may precede flowering as in gladiolus or flowering occurs before the leaf emerges e.g. Amaryllis, *Haemanthus* species. After growth and flowering, the plants mostly enter into a rest period. The duration of dormancy varies with the type of plants and environmental factors such as temperature and humidity. Bulbous flowers such as Gladiolus, Lilium, Alstroemeria, Tulip, Freesia are in the list of the top ten cut flowers traded in the international market. Most of these flowers require low temperature and are commercially popular in North India.

Bulbous plants display grandure throughout the year. Some of them possess very attractive foliage while others are valued for their fragrance. But above all, their flowering spikes are the centre of attraction. They offer a wide variety of forms and colours, from bright primary shades to delicate pastel hues. Adaptation to tough environments makes bulbous plants invaluable in any garden.

CLASSIFICATION OF BULBOUS FLOWERING PLANTS

Classification on the Basis of Storage Structures

- **Bulbs:** Amaryllis, Dutch Iris, Hyacinth, Lilies, Tulips, Narcissus, Daffodil, Tuberose, Zephyranthus, Crinum

- **Tubers**: Caladium, Calla Lilies, Dahlia
- **Tuberous roots**: Anemones, Gloxinia, Ranunchulus
- **Rhizomes**: Agapanthus, Alstroemeria, Iris, Canna
- **Corms:** Gladiolus, freesia

Classification on the Basis of Plant Height

- **Dwarf bulbous plants:** Tulip, Crocus, Hyacinthus, *Iris reticulata, Muscari botryoides*
- **Medium tall bulbous plants:** Fritillaria, Iris, Narcissus, Paeonia
- **Tall bulbous plants:** Agapanthus, Belamcanda, Hemerocallis, Kniphofia

Classification on the Basis of Ornamental Value

- **For bedding purpose:** Canna, Dahlia, Crinum and Tuberose
- **For informal garden:** Amaryllis and Gladiolus
- **For rock garden:** Amaryllis, Alliums and Crinum Anemone, Anthericum, Bulbocodium, Chionodoxa, Cyclamen, Eranthis, Erythronium, Galanthus, Ipheion, Muscari, Ornithogalum, Oxalis, Romulea, Rhodohypoxis and Scilla
- **For pots and hanging baskets:** Pelargonium, Freesia, Gladiolus, Tuberose and Achimens
- **For glass house:** Achimens, Begonia and Caladium
- **For shady areas:** *Crinum, Allium ursinum, Anemone blanda, Anemone nemorosa, Arum italicum, Convallaria majalis, Corydalis flexuosa, Cyclamen purpurascens, Disporum flavescens, Erythronium, Fritillaria pallidiflora, Galanthus, Hyacinthoides* non-scripta, *Iris douglasiana, Leucojum vernum, Lilium martagon, Ranunculus ficaria, Sanguinaria canadensis, Smilacina racemosa,* Trillium and *Uvularia grandiflora*
- **For indoor garden:** Begonia, Amaryllis, Gloxinia and Caladium

Classification on the Basis of Flowering Season

The bulbous plants flowering in different seasons such as spring, summer, and autumn add an extra layer of colour to the garden. These can be planted under shrubs or between perennials in the flower border or can be naturalized informally in lawns and orchards, or formally in seasonal bedding display.

- **Spring flowering bulbs:** Allium, Arum, Asphodelus, Camassia, Convallaria, Crocus, Cyclamen, Eranthis, Freesia, Fritillaria, Galanthus, Hyacinthus, Hippeastrum, Iris, Ixia, Leucojum, Muscari, Narcissus, Ornithogalum, Ranunculus, Scilla, Trillium, Tulipa and Zephyranthes.
- **Summer flowering bulbs:** Achimenes, Agapanthus, Allium, Alstroemeria, Amaryllis, Anemone, Begonia, Calochortus, Canna, Crinum, Crocosmia, Dahlia, Dierama, Eucomis, Galtonia, Gladiolus, Gloriosa, Haemanthus, Hymenocallis, Lilium, Oxalis, Pancratium, Paradisea, Polianthes, Sprekelia, Tritonia, Watsonia, Zantedeschia, etc.
- **Bulbs flowering in the fall season:** *Crocus,* Colchicum, Cyclamen, Lycoris, Nerine, Sternbergia.
- **Winter flowering bulbs:** *Galanthus, Crocus, Cyclamen* and *Eranthis*

The major bulbous flower crops for hilly areas are Gladiolus, Zantedeschia, Lilium, Dahlia, Tulip, Alstroemeria, Tuberose, Rhizomatous iris, Freesia, Canna, Narscissus, Liatris etc. The minor bulbous flower crops for hilly areas are Grape hyacinth, Mexican sunflower, Watsonia, Tigridia, Crocus etc.

Bulbous flowering and foliage plants grown in tropical and subtropical areas may also be grown in low hilly areas and these include Achimenes, Alstroemeria, Amaryllis, Canna, Cooperia, Crinum, Dahlia, Eucharis, Eurycles, Gladiolus, Gloriosa, Haemanthus, Hedychium, Hemerocallis, Hippeastrum, Hymenocallis, Nelumbo, Pancratium, Polyanthes, Gloxinia (Sinningia), Zantedeschia, Alocasia, Asparagus, Caladium, Calathea, Kaemferia, and Sansevieria .

Bulbous plants, which grow well in the high hills are Agapanthus, Allium, Anemone, Crocus, Cyclamen, Freesia, Fritillaria, Iris, Ixia, Hyacinthus, Lilium, Narcissus, Nerine, Paeonia, Kniphofia, Ornithogalum, Ranunchulus, Tuberous begonia, Tulip, Watsonia, Sparaxis etc.

Bulbous plants are usually herbaceous perennials, which play an important role in the beautification of our landscape. These plants can be used in formal as well as informal landscape in beds, borders, pots etc. Herbaceous border comprising of annuals and perennials can be planned with the help of bulbous plants, in both formal and informal situations where they provide a succession of bloom throughout the year (Jindal 1968). Dwarf species of bulbous plants such as Tulip, Crocus, Hyacinth, *Iris reticulata and Muscari botryoides* can be planted in beds in combination with dwarf annuals such as Pansy, Anchusa, Candy tuft, Linaria, Sweet alyssum and Forget-me-not. This arrangement provides bloom from March to June. Medium sized bulbous plants such as

Fritillaria, Iris, Narcissus and Paeonia can also be planted at the rear of dwarf seasonal beds to flower from spring to early summer. Colour can be added to bulbous plants during summer by planting tall bulbous plants such as Agapanthus, Belamcanda, Hemerocallis and Kniphofia behind the beds planted with medium sized summer annuals such as Cockscomb, French marigold, China aster, Bedding dahlia, Gomphrena, Balsam and Salvia (John et al 2007).

MAJOR BULBOUS FLOWER CROPS GROWN IN HILLY AREAS

Gladiolus (*Gladiolus hybridus* Hort.)

Origin: South Africa

Family: Iridaceae

Gladiolus is a flower of breathtaking beauty with the widest range of colours, sizes and forms of flowers. The cultivation of gladiolus is important in hilly areas as it is an off-season crop grown in plains of India. It is being grown commercially in almost all hilly areas of Himachal Pradesh, Uttarakhand, Jammu and Kashmir, North Eastern states. It grows in sunny conditions. An average temperature of 15-20°C is favourable for better growth of the plant. It prefers sandy loam soil with a pH of 5.5 to 6.5. The varieties are Aldebaron, Jester, Nova Lux, Top Brass, Topaz, Vink's Glory, Amercian White, White Friendship, White Prosperity, Pink Friendship, Suchitra, Spic and Span, Amercian Beauty, Oscar, Red Majesty, Blue Lilac, Red Beauty, Interpid, Trader Horn, Mayur, Marvellous. The varieties evolved by Indian Agrucultural Research Institute (IARI), New Delhi are Agni Rekha, Gulal, Gunjan, Jyotsna, Kamini, Pusa Suhagin, Urmi, Urvashi, Suryakiran, Suchitra etc. and the varieties evolved by Indian Institue of Horticultural Research (IIHR), Bengaluru are Aarti, Apsara, Arka Kesar, Arka Suvarna, Darshan. Dhiraj, Kum Kum, Poonam, Shobha, Sindhur (Janakiram et al 2007). The planting and flowering time in low hills is July-October and November-March respectively. In mid hills, it is February-March and May-October respectively, and in high hills it is April-May and July-November respectively. The recommended spacing is 30 x 30 cm for better growth of plants. Planting of corms is done 8 cm deep in soil (full size i.e. 4 cm and above). Planting depth should be slightly shallower on heavy soils compared to that on lighter soils. Planting stock is propagated through cormels and flowering stock through corms (size 4 cm and above) (Panwar and Saxena 2001).

Dormancy of corms can be broken by treatment with 150 ppm ethrel solution for 24 hours or a combination of ethrel (150 ppm) treatment and storage of corms at 5°C for a week. This treatment is effective for earlier sprouting and flowering. The spikes should be cut at tight bud stage with two to three leaves remaining on the stem when the lowest bud shows colour. For local market, the spikes should be cut when the lower floret starts opening. The spikes should be harvested during morning or evening hours.

Lilium (*Lilium* sp.)

Origin: Northern Hemisphere up to South Canada and Siberia, Florida, and Nilgiri mountains of India

Family: Liliaceae

Lilium is one of the most important bulbous crops which is beautiful, attractive and has bright flowers with long vase life. They fetch higher market price than other commercial flowers.

Lilium can be grown in almost all kinds of soil. However, sandy loam soil with pH range of 6.0-7.0 is ideal for cultivation. For better growth, the temperature should be 15-25°C during day and 8-10°C during night. It grows well under low light intensity. Therefore, 75% shade nets should be used during the summer season and 50% shade nets should be used during winter season.

Lilium is propagated using seeds, scales, bulblets and bulbils, micropropagation etc. The varieties of Asiatic Lily are Alaska, Beatrix, Connecticut King, Cordelia, Elite, Paris, London, Geneva, Gran Paradiso, Menton, Massa etc. whereas Star Gazer (Pink), Star Gazer (White) are common in Oriental Lily. In low hill areas, planting period is from October-November, which gives flowers from January-February. In medium and high hills, planting is done from April–May, which gives flowers from July-August. To plant Asiatic and Oriental Lily, bulbs of bulbs of size 10-12 cm and 14-16 cm respectively should be used. The plants should be staked during their growing period especially of those cultivars, which have more than 80 cm height. Excess water is harmful to the crop as it checks aeration in root zone. Flowers become ready for harvesting between 90-120 days after planting and should be harvested during morning and evening hours.

Zantedeschia (*Zantedeschia aethiopica*)

Origin: Southern Africa

Family: Araceae

Zantedeschia is a genus of herbaceous flowering plants. The name of the genus was given as a tribute to Italian physician and botanist Giovanni Zantedeschi (1773–1846) by the German botanist Kurt Sprengel (1766–1833).

The *Zantedeschia* species are poisonous due to the presence of calcium oxalate. All parts of the plant are toxic and produce irritation and swelling of the mouth and throat, acute vomiting and diarrhea. The most common species was long known in the flower trade as *Calla aethiopica* which was changed to *Richardia africana*, but presently it is renamed as *Zantedeschia aethiopica*. It is a robust, evergreen to deciduous, erect and clump-forming plant with thick rhizomes and white fleshy roots. The leaves are large, leathery, glossy dark green with lighter green fine veins and have a laminae-sagittate or ovate-cordate form. The stout, green scape is of similar height as of the leaves. The flower has a very distinctive structure which appears to be a large petal but actually it is a modified leaf called 'spathe' of ivory-white colour with bright green base on the outer side. The spathe is up to 25 cm long, funnel-shaped, narrowing towards the tip with a small recurved apex; inside it contains the actual flower which is 10-12 cm long spadix of bright yellow colour having a basal female zone of ¼ - ½ of the spadix continued with an upper male zone.

The common varieties are Striped Hybrida, Gigantea, Green Goddess, Childsiana, Godfregana, Candidissima, Grandiflora, Devonensis and Little Gem. The plants grow well in open field conditions and plenty of sunlight. The temperature requirement varies from 10°C to 25°C. However, these plants cannot tolerate frost and severe winter. For optimum growth, the plants require soil, rich in organic matter, and its flowering requires sufficient moisture in the substrate. The plant has an interesting mechanism which allows it to grow in wet conditions i.e. *"The leaves of the arum are very interesting in that they contain water stomata which can discharge excess water, by a process known as '**guttation**'. This prevents waterlogging and enables arum lilies to grow in wet conditions"* Alice Aubrey (2001).

The best quality and most uniform culture requires special attention to media selection, water management and forcing temperatures, as well as a strong preventive fungicide programme. Disinfected tubers should be planted directly in raised beds at a spacing of 30-60 cm. Planting should not be too close as it leads to lower flower production caused by lack of light per plant. The planting depth, depending on the tuber size, is approximately 10 cm and planting density is 30 rhizomes per square metre. Shallow planting is not recommended as the root system will develop close to the soil surface. The rhizomes take 13-16 weeks to start flowering. The flowers should be harvested early in the morning or late in the evening when the temperature is not too high and they should be placed in the cooling cell immediately. The flowers are harvested by pulling them up (De Hertogh and Le Nard 1993).

Canna (*Canna* species)

Origin: Tropical and subtropical regions of the New World

Family: Cannaceae

Canna (Canna lily) plants are large tropical and subtropical perennial herbs with a rhizomatous habit. The broad, flat, alternate leaves, a typical feature of this plant, grow out of a stem in a long narrow roll and then unfurl. The leaves are typically solid green but some cultivars have brownish, maroon, or even variegated leaves. The flowers are composed of three sepals and three petals, and they are small and hidden under extravagant stamens. Canna grows from swollen underground stems i.e. rhizomes.

All parts of the plant have commercial value: rhizomes for starch, which is consumed by humans and livestock, stems and foliage for animal fodder, young shoots as a vegetable and young seeds as addition to tortillas. They can be used in herbaceous borders, tropical plantings, and as a patio or decking plant. The varieties are *Canna* 'Theresa Blakey' (Crozy Group), 'Edulis Dark' (Agriculture Group), 'Oberon' (Miniature Group), etc. Canna may be propagated sexually i.e. through seeds and asexually i.e. through rhizome division,

Cannas grow best in full Sun with moderate water in well-drained, rich or sandy soil. Cannas grow from perennial rhizomes but are frequently grown as annuals in temperate zones for an exotic or tropical look. The rhizomes are frost tender and will rot if left unprotected in freezing conditions. In areas where temperature goes up to -10°C in the winter, the rhizomes can be dug up before freezing and stored in a protected area (above 7°C / 45°F) for replanting in the spring. Otherwise, it is recommended that Cannas are protected by a thick layer of mulch over winter (Motial 1982).

Dahlia (*Dahlia variabillis* L.)

Origin: Mexico, Central America, and Colombia

Family: Compositae (Asteraceae)

Dahlia is one of the most important perennial garden plants. The Abbe Canailles gave the name Dahlia in 1791 after Swedish botanist Anders Dahl (Smith 1963). It was originally discovered by Aztecs who called it cocoxochiti. Dahlia was introduced in India as early as 1857 under the auspice of the Agri-Horticultural Society of India, Kolkata.

Dahlia is used as cut flower, pot plant for decoration purpose, garland making etc. Dahlias are easy to grow both in fields and in pots, and are extensively used for home decoration. Dahlias were classified and published by the National Dahlia Society of England (NDS). These classes are Single flowered Dahlia, Star Dahlia, Anemone, Collerette, Paeony flowered, Decorative (Giant - more than 25 cm diameter; Large flowered - 20-25 cm; Medium flowered - 15-20 cm; Small flowered - 10-15 cm; Miniature flowered - less than 10 cm), Pompon Dahlia, Cactus Dahlia,

Ball Dahlia, Miscellaneous Dahlia (Modern Pompon, Orchid Dahlias, Chrysanthemum flowered Dahlia, Fimbriated, Rose flowered, Miniature Collerette, Needle pointed Dahlias), Water lily dahlia and Star-flowered.

The modern garden cultivars of Dahlia have developed as a result of crossing between various species and varieties such as *Dahlia variabilis, Dahlia coccinea, Dahlia crocea, Dahlia imperialies, Dahlia juarezii and Dahlia merckii.* These varieties are African Queen, Autumn Tips, Axford Gold, Art Link Letter, Bilis Whit, Challenger, Croydon Apricot, Croydon Delicate, Croydon Lilac, Croydon Masterpiece, Croydon Monarch, Gracy Hey, Majestat, Our Friend, The Dragon, Uchuu, Ami Pardieu, Arab Queen, Capistrano, Nita and Silver Wedding.

Dahlias are mainly propagated by seeds, division of tuberous roots and by terminal stem cutting, grafting and micropropagation. It grows well in any type of rich and porous soil but generally prefer well drained, deep, fertile and moist soil with pH 6.5. An open and sunny place but sheltered from exposure of strong wind is ideal. A frost free atmosphere is also necessary for proper growth. The tuberous nature of Dahlia enables them to survive in the dormancy. So, the gardeners in temperate climates with frosts can grow dahlias successfully, provided that the tubers are lifted from the ground and stored in cool yet frost-free conditions during the winter. In the plains the tubers are generally planted from September-October and in hills the tubers are planted in April. The planting distance varies according to the type i.e. 75 cm in case of tall, large flowered types, 60 cm for medium sized plants and 30-45 cm in case of dwarf cultivars. Dahlias are benefited by heavy watering at longer spells rather than by frequent light watering. Disbudding, pinching, mulching, staking and pruning are important operations in dahlia for better growth (Vinayananda 1986).

Alstroemeria (*Alstroemeria* sp.)

Origin: South America (Chile, Brazil, Peru, Paraguay, Venezuela and Argentina)

Family: Alstroemeriaceae

Alstroemeria is commonly known as Brazilian Parrot lily or Incas lily or Peruvian lily. It is popularly grown for cut flower and is listed among top ten international cut flowers. Alstroemeria is propagated by traditional method of rhizome division preferably during dormancy. Commercially, micropropagation is the most popular method of multiplication in which 4-7 fold increase occur after every four weeks. It can also be propagated by seeds which should be sown immediately after pod ripening.

Alstroemeria is sensitive to light and temperature. It prefers cool climate with partial shade. The control of flowering process requires thermophase and photophase. The requirement of thermophase must be fulfilled before photophase. The optimum temperature in greenhouse during night and day is 15°C and 18°C respectively. The newly planted rhizomes/plants should not get more than 13 hrs light a day and that too for at least 6-8 weeks, which will allow the roots to develop sufficiently before flowering (Raj 2003). After this, supplementary light of more than 16 hrs a day will ensure early, profuse flowering for longer duration. Poor light leads to blind shoots and flower blasting, but excessively long light duration could also lead to production of smaller number of shoots. Short light days suppress rhizome formation and shoot development. Relative humidity inside the greenhouse should be 60-70% throughout the growing season. The CO_2 concentration up to 1200 ppm is always beneficial in increasing flower yield especially at higher temperature. Cool and well drained soil rich in organic matter is good for Alstroemeria. It can be grown on nearly all soil types. In general, the optimum soil pH is 6.0-7.0. The ideal growing media has one part each of sphagnum peat moss, soil and sand, perlite, expanded clay and gravel. In peat soils, Alstroemeria shows a much quicker and luxuriant growth than in clay soil. The time of planting depends upon the type of variety, the climate, circumstances, experience of the grower etc. Generally, beds should be 15-20 cm deep so that roots grow well during a three-four year production cycle. The growing point of rhizomes is planted 7-10 cm deep. The spacing varies with variety and purpose whether cultivation is for cut flower or for planting material production or both. The optimum spacing between plants and rows should be 40- 50 cm. The planting time in North Indian hills is from September-October or February-March and in South India it is from May-June or November-December. Irrigation depends on the prevailing weather conditions. Moist soil is beneficial for better growth, flowering and development of rhizomes for which irrigation at 7-10 days interval is ideal. Mulching, support and shading up to 30% are important operations for healthy flowers.

Tulip (*Tulipa* sp.)

Origin: Southern Europe, North Africa, and Anatolia

Family: Liliaceae

Tulips were first introduced in the Netherlands in the late 1500's. Tulips are spring-blooming perennials that grow from bulbs and are used as cut flower in international market. Most tulips produce only one flower per stem, but a few species bear multiple flowers on their scapes (e.g. *Tulipa turkestanica*). The showy, generally cup or star-shaped tulip flower has three petals and three sepals, which are often termed tepals because they are nearly identical. These six tepals are often marked near the base with darker colour. Tulip flowers come in a wide variety of colours, except pure blue. The flowers have six distinct, basifixed stamens with filaments shorter than the tepals. Tulip stems have few leaves, with larger species tending to have multiple leaves.

The early flowering tulips include varieties such as Duc Van Tol, Single Early and Double early. The mid season flowering tulips include varieties such as Mendel, Triumph and Darwin hybrid and the late season flowering varieties are Darwin, Single late, Parrot flowered, Fringed, Double late, Lily flowered and Viridifolia. (Bhalla and Sullhan 2003). The common species are *Tulipa biflora, Tulipa festeriana, Tulipa chrysantha, Tulipa montana, Tulipa clusiana, Tulipa haferi, Tulipa persica, Tulipa prastans, Tulipa tarda, Tulipa greigii* and *Tulipa eicheri.*

These are propagated through bulbs or seeds. Tulips are indigenous to mountainous areas with temperate climates and need a period of cool dormancy, known as vernalization. They thrive in climates with long, cool springs and dry summers such as of Himachal Pradesh and Kashmir. The soil should be well drained, with pH 6-7, and low in soluble salts. The pH of the soil should not be acidic, as it reduces the effectiveness of ancymidol. The soil temperature should be 13°C. The bulbs of size 10-12 cm should be planted with their nose about 2-3 cm deep in the planting medium. Planting time in Northern India is from mid September to mid December. In general, planting density of 50-68 bulbs per square metre is used. Ground or raised beds may be used with spacing of 15×10, 15×15 or 10×10 cm.

Narcissus/ Daffodils (*Narcissus* sp.)

Origin: Europe, North Africa and Asia

Family: Amaryllidaceae

The Narcissus flower is perceived in the West as a symbol of vanity, and in the East as a symbol of wealth and good fortune. All *Narcissus* species have a central bell bowl or disc-shaped corona surrounded by a ring of six floral leaves called the perianth, which is united into a tube at the forward edge of the three-locular ovary. The seeds are black, round and swollen with a hard coat. The three outer segments are sepals, and the three inner segments are petals.

The species of wild Narcissi, normally which are highly fragrant, show a wide range of shapes and sizes than the garden cultivars. Some of the species are *Narcissus alcaracensis, Narcissus assoanus, Narcissus asturiensis, Narcissus bugei, Narcissus jonquilla* (Jonquil), *Narcissus longispathu, Narcissus papyraceus* (Paperwhite Daffodil), *Narcissus poeticus* (Poet's Narcissus), *Narcissus pseudonarcissus* (Wild Daffodil), *Narcissus radinganorum, Narcissus romieuxii, Narcissus tazetta, Narcissus triandrus* (Angel's Tears) and *Narcissus medioluteus*. The popular varieties grown worldwide are Barrett Browning (White orange), Fortissimo (White yellow), Golden Torch (Deep Golden Yellow), Irish

Luck (Golden yellow), King Alfred (Yellow), Ice Follies (White) Las Vegas (White Yellow), Minnow (Yellow), Moonrise (Lemon yellow), Mount Hood (White) Salome (White Pink Yellow) Scarlet Elegance (Yellow with rich Orange Red cup), and Tahiti (Yellow red) (Misra *et al.* 2003).

Narcissi are propagated sexually by seed and asexually by bulbs, bulb offsets or divisions. Narcissi are not very demanding in their soil requirement but good results are obtained on well drained soil rich in humus, which has a moderate water holding capacity. The pH of the soil should range from 6.0-6.5, although they can be grown successfully at a pH as low as 4.9. Lime is usually not added to soil when growing Narcissi. It performs well under shady conditions. Longer photoperiods only promote the length of the stem. After bud initiation, it requires low temperature. Planting of narcissus bulbs is usually rotated every 2-3 years to avoid diseases. Planting of bulbs is done in beds (1 m width) at a spacing of 10-15 cm × 30-40 cm between bulbs and rows respectively. These may be planted any time after the bulbs have been lifted, but planting is usually done from mid August and continues till mid September. Narcissus requires copious watering.

The American Daffodil Society (ADS) divides all *Narcissus* into 16 horticultural divisions, based partly upon flower form and partly upon genetic background. The divisions include Trumpet Daffodils, Large-cupped Daffodils, Small-cupped Daffodils, Double Daffodils, Triandrus Daffodils, Cyclamineus Daffodils, Jonquilla Daffodils. Tazetta (Poetaz or Bunch-flowered) Daffodils, Poeticus (Poet's) Daffodils, Bulbocodium Daffodils, Split-corona Daffodils, Collar Daffodils, Papillon Daffodils, Miscellaneous Daffodils, Wild Variants and Wild Hybrids, and Miniature Daffodils.

Iris (*Iris* species)

Origin: South East Asia and Southern Europe

Family: Iridaceae

The genus Iris is widely distributed throughout the North Temperate Zone. Irises are perennial herbs which mostly thrive best in the hills. The Irises have two forms of growth i.e. having fleshy rhizomes and those having bulbs. These are again sub-divided into 10 or 11 groups and the largest group is the bearded Irises. Bearded Irises are of two types i.e. dwarf bearded Irises (developed from *Iris chamaeiris*, *Iris flavescens* and *Iris pumila*) and tall bearded Irises (*Iris germanica*). The tall bearded Irises can be grown in sunny conditions and a well drained soil containing lime. These are propagated from seeds and by rhizome or bulb division. The Himalayan Irises, *Iris nepalensis, Iris kashmiriana* and *Iris kamaonensis* are rhizomatous and can be grown from low hills to high hills.

MINOR BULBOUS ORNAMENTAL PLANTS FOR HILLY AREAS

Fragrant bugle-lily (*Watsonia marginata*)

Origin: South Africa

Family: Iridaceae

It is grown in gardens for its showy spikes of flowers. It is a perennial herb growing from a corm. It grows up to a maximum height well over one metre while flowering; sometimes it may grow up to two metres. Each corm produces three or four erect leaves, which measure up to 80 cm long and 5 cm wide. The leaves are blue-green with thickened yellow margins. The inflorescence is a dense spike of 30-50 flowers, which can have any shade of pink or sometimes dark red or white. The flower is actinomorphic or radially symmetrical, unlike those of other *Watsonia*, which are zygomorphic.

Pink watsonia (*Watsonia borbonica*)

Origin: South Africa

Family: Iridaceae

Watsonia borbonica is a tender to half-hardy herbaceous perennial bulbous plant with tall spikes of rose-pink, trumpet-shaped flowers, which makes a picturesque display. It flowers for up to 4 or 5 weeks. It is a beautiful garden subject that needs little maintenance. The genus *Watsonia* was named by Philip Miller of Chelsea after his friend Sir William Watson (1715-1787), a London physician and naturalist.

It is deciduous, growing during autumn-winter-spring and dying back after flowering in spring, and remaining dormant during summer. *Watsonia borbonica* produces abundant offsets (daughter corms), and hence the easiest method of propagation is by division. Clumps can be lifted and divided at any time during the dormant season (December-February) and re-planted in spring (March-April). Corms should be planted at a depth of approximately half of their diameter. It is hardy and easy to grow, with a long flowering period. Although adapted to a winter rainfall climate, it does best in full sun, and in well drained, well composted soil. It will thrive under summer rainfall conditions provided it is grown in well drained soil. It can withstand mild winter frost but should be lifted, dried and stored during winter. After flowering, the leaves and stems can be cut back. Commonly, the best planting season is between the late summers and the early autumn. However, planting is recommended in spring in the cold region. During the growing period, wait until the surface of the soil is dry, and then water thoroughly. Stop watering if the flowers have faded.

Crocus (*Crocus sativus*)

Origin: Asia Minor (Italy to Kurdistan)

Family: Iridaceae

Crocus can be grown well only in the hills. It is a winter-flowering species of dwarf habit and grass like foliage, which bears small rounded fragrant flowers on long slender stalks. It flowers from September to December. Commonly known as 'Saffron', this plant is used as an ornamental plant, spice, food colorant and less extensively, as a textile dye or perfume. However, due to its analgesic and sedative properties, folk herbal medicines have used saffron for the treatment of some illnesses for centuries (Basker and Negbi 1983).

Crocus is most impressive when planted in mass. *Crocus speciosus* is the only species, commonly called Autumn Crocus, Fall Crocus or Ornamental Crocus, which is grown in the hills. The varieties are Flower Record (purple), Jeanne d'Arc (white), Mammoth Yellow (yellow), Pickwick (white with blue stripes), and Remembrance (lavender). Crocuses are grown in rockeries, beds, pots, in small clumps under trees, shrubberies and lawns. The time of planting for the autumn-flowering crocuses is from July-August and for spring-flowering, it is from October-November. The corms are planted very closely to each other and can be grown in sunny and semi-shady places.

Grape Hyacinths (*Muscari armeniacum*)

Origin: Southeastern Europe

Family: Liliaceae

Grape hyacinth is a mid-spring-blooming, perennial bulb in the lily family. Grape hyacinth's common name refers to the plant's clusters of small, bell-shaped, cobalt-blue flowers (with narrow, white rims) that look like clusters of grapes. Grape hyacinth grows four to eight inches tall, and produces floppy, narrow, green leaves that emerge from the ground in early spring. Each bulb produces one to three flower stalks with 20 to 40 tightly packed flowers per stalk. After flowering in early summer, the plant's foliage dies. However, unlike many other spring-blooming bulbs, grape hyacinth starts growing actively again in mid-fall and persists through the winter.

The varieties of grape hyacinth are - 'Album' which is white-flowered and fragrant, but not as vigorous as the standard type; 'Blue Spike' is flax blue with double flowers, and blooms slightly later than the standard type; 'Fantasy Creation' has double blue flowers that may develop green overtones as the flowers ages; 'Saffier' has sterile, blue flowers that are longer lasting than most varieties; 'Superstar' has densely packed white-edged, periwinkle-blue florets, topped with a cap of paler florets.

The Grape Hyacinth should be planted in the fall season in well drained soil. It tolerates both sunny and shady conditions. Plants in full Sun tend to have better vigour, however, those planted in partial shade tend to flower for a longer period of time. When planting bulbs, place them three to four inches deep and two inches apart. Use bone meal at the time of planting, as well as after blooming, to improve grape hyacinth performance. Be sure that plants are well-watered during leaf growth and flower production, but watering should be reduced after the foliage begins to die back.

The Tiger-flowers (*Tigridia* sp.)

Origin: Southeastern Europe

Family: Iridaceae

Tigridia have large showy flowers and its one species, *Tigridia pavonia* (Linnaeus fil.) de Candolle, is often cultivated for flowers. Approximately, 35 species in this family grow in the Western Hemisphere, from Mexico to Chile. The Tigridia flower is short lived, mostly blooming only for a day, but often several flowers will bloom from the same stalk. Usually they are dormant during the dry season of winter. Its roots were eaten by the Aztecs of Mexico. They termed the roots cacomitl and its flower 'Jaguar flower'. The genus name means 'tiger-like' and alludes to the colour and spotting of the flowers of the *Tigridia pavonia*. The other species of this genus are *Tigridia chiapensis, T. dugesii, T. durangense, T. ehrenbergii, T. hallbergii, T. mexicana, T. pavonia, T. multiflora, T. orthantha, T. vanhouttei, T. violacea, T. flammea, T. immaculata, T. inusitata, T. philippiana, T. huajuapanensis, T. pasiflora, T. meleagris, T. molseediana, T. alpestris, T. seleriana* and *T. bicolor*.

Amaryllis (*Amaryllis belladona*)

Origin: South Africa

Family: Amaryllidaceae

It is a plant with strap-shaped, glossy, green leaves that produces a few large funnel-shaped flowers on a stout stalk, from March-May. Leaves appear at the end of the flowering season and die out before the winter sets in. In warm humid region, the plants retain leaves throughout the year and bear flowers during the season. Offsets, developed from the original bulb, are separated for multiplication. Bulbs are planted 5-8 cm deep in soil, from December-January. The plants grow well in rich and porous soil. Amaryllis is not disturbed for several years from where it is planted and the plants continue to produce new bulbs and flowers. It is very suitable for planting in border, shrubbery and in pots. Although rose or rosy pink colour is common; white to purplish colour is also planted (Randhawa and Mukhopadhya 2004).

Gloriosa (*Gloriosa* sp.)

Origin: Africa and tropical Asia

Family: Liliaceae

As the name implies, the genus *Gloriosa* means 'full of glory', popularly known as climbing or creeping lily. They are tall creeping plants that support themselves by means of tendrils, which arise from leaves. This genus comprises of about six species of rhizomatous plants. The commonly grown species in India are *Gloriosa superba* and *Gloriosa rothschil-diana*. Leaves are oblong, lanceolate. Flowers are showy on long pedicels in leaf axils, perianth of six distinct long segments; six stamens with versatile anthers.

Well drained soil in a sunny location is ideal for planting Gloriosa and the attractive flowers make a fine display of colour when trained on bamboo framework or low trellis. Gloriosa is vegetatively propagated from rhizome, which may be cut in pieces and planted at a depth of 3-4 cm in April-May. The plant flowers from July-August. After flowering, the plant begins to wither and rhizome becomes dormant. If left undisturbed, Gloriosa continues to produce flowers for several years in the same place.

Achimenes (*Acimenes longiflora*)

Origin: Guatemala, Mexico

Family: Gesneraceae

It is a dwarf plant with tubular petunia-like flowers of violet-blue colour, which are whitish underneath. It flowers from June-November. The propagation is done by tuber or terminal cuttings raised during the rains. A few criteria for successful cultivation are adequate drainage, plenty of moisture and partial shade.

Scented Gladiolus (*Acidenthera bicolor*)

Origin: Abyssinia

Family: Iridaceae

It is a tall plant, which resembles gladiolus and bears sweet scented white flowers with star–shaped, crimson-maroon blotch in the throat. The propagation is done by corms. In hills, planting is done from March-April for flowering in autumn. A well drained richly manured soil and sunny sheltered location is suitable for its cultivation.

Blue African lily (*Agapanthus umbellatus*)

Origin: Africa

Family: Liliaceae

The plant has graceful evergreen leaves, funnel-shaped pale blue or blue flowers, which are borne in umbels of 10-30 flowers. A sunny location, rich soil and plenty of water is good for its cultivation. Planting is done in spring for flowering, from July – September.

Common Hyacinth (*Hyacinthus orientalis*)

Origin: Syria, Lebanon, Turkey Palestine and Asia Minor

Family: Liliaceae

It produces slender flower stems bearing racemes of small blue florets. Now a days, hybrids of different colours from white to yellow, purple, mauve, dark and light blue, pink and red are available. It flowers well only in the hills. Propagation is done by bulbs and bulblets and sometimes using seeds. These are planted in the month of February and they flower from May to June.

Torch Lily (*Tritoma uvaria*)

Origin: South Africa

Family: Liliaceae

The plant is tall and has reed like foliage in clusters. The flowers are borne at the end of a leaf less stem in densely-packed racemes. It is propagated through seeds or rhizomes. It is planted during February-March and flowers during rainy season.

Belladonna Lily (*Amaryllis belladonna*)

Origin: South Africa

Family: Amaryllidaceae

The plant is tall having long strapped leaves. It has large, trumpet-shaped, rose coloured flowers, which are borne in umbels of 2-4 flowers at the end of a long, hollow but stout stem. Propagation is done by bulbs. Bulbs are hardy and bloom from winter to summer in the hilly areas. The planting time for bulbs is from March to April.

Wind flower, anemone (*Anemone coronaria*)

Origin: South Europe to Central Asia

Family: Ranunculaceae

The plant has solitary single or semi double large flowers of crimson, scarlet, purple, pink, blue, mauve or white colour. These plants grow well on high hills. These are raised from seeds, division and cuttings taken from rootstock. A well drained soil rich in manure, and sunny or semi-shaded conditions are the best for its cultivation. The planting time is from autumn or spring in the hills.

Black berry lily (*Belamcanda chinensis*)

Origin: China

Family: Iridaceae

It is a rhizomatous plant with Iris like leaves and orange coloured flowers with crimson spots which are borne on long stems. The plants are suitable for pots, shrubberies and borders. It is propagated from seeds or by division. Planting is done during October-November in hills for flowering during August- September.

Kaffir lily (*Clivia miniata*)

Origin: Natal, Africa

Family: Amaryllidaceae

It is an evergreen flowering bulbous plant suitable for the hills only. Planting is done in the sunny and semi-shaded places from February-March. The orange-yellow or scarlet coloured funnel-shaped flowers are borne in umbels.

Lily of the valley (*Convallaria majalis*)

Origin: Asia and Europe

Family: Liliaceae

Dwarf plant with creeping rhizomes and bell-shaped flowers are very popular for their fragrance. The creeping rootstock is planted horizontally about 3 cm below the surface. The planting time is autumn or spring and it flowers during May-June.

Sow broad or Cyclamen (*Cyclamen* sp.)

Origin: Mediterranean region

Family: Primulaceae

These are herbaceous, low growing plants with tubers or corms. The commonly grown species is *Cyclamen persicum* var. *giganteum*, which has large flowers of different colours such as white, pink, purple and red. Propagation is done from seeds but corms or tubers are more convenient. It likes humidity and can be maintained by frequent syringing of the foliage. The plants need partial shade, moist conditions and a temperature ranging from 10-15°C to produce best quality flowers. The corms are planted in October in the hills or during August for indoor display from winter to early summer.

Crown Imperial (*Fritillaria imperialis*)

Origin: Asia, Europe and North America

Family: Liliaceae

The plants are tall with narrow, whorled foliage, brick red, coppery red, yellow and orange flowers having an obnoxious foxy odour. Propagation is done through bulbs, which are planted between August and November. Partial shade, rich and moist soil is important for successful cultivation.

Day lily (*Hemerocallis fulva*)

Origin: Asia and Europe

Family: Liliaceae

This is a tall hardy perennial with grass like leaves and yellow-orange flowers. The large and trumpet-shaped flowers are borne in long and branched stems. These are suitable for mid hills. Plants are propagated by division during March. These prefer rich soil, sunny conditions and abundance of moisture. The flowering season is between April and June (Randhawa and Mukhopadhya, 2004).

Plantain lily (*Hosta plantaginea*)

Origin: China and Japan

Family: Liliaceae

The plants of *Hosta* species are perennial to the hills. These are particularly suitable for planting in a wild garden and also in shrubberies, on slopes of land, rockeries and pots. Plants are propagated by division of rhizomes. The flowers are trumpet-shaped, white and fragrant.

Star of Bethlehem (*Ornithogalum umbellatum*)

Origin: Europe including Britain and North Africa

Family: Liliaceae

Plants produce star-shaped and white coloured flowers, with green stripes on the outer side, which are borne at the end of a scape in racemes of 12-20 flowers. It is propagated from offsets or bulblets. A rich and well drained soil and an open situation are ideal for its growth but it can also tolerate partial shade. In hills, plants should be protected from frost. In the hills planting is done during October and it flowers from April to June (Randhawa and Mukhopadhya 2004).

Table 1. Bulbous flower crops

Scientific name	Common name	Mode of propagation	Planting time	Plant height	Flowering time	Bloom colour
Agapanthus umbellatus	African lily	Bulb/seed	November and March	90-100	July-August	Blue-violet, white
Aistroemeria spp.	Peruvian Lily	Rhizome	November and March	50-75	April-October	Varying flower colours
Amaryllis belladonna	Belladonna Lily	Bulb	June-July	50-60	August-September	Rose
Anemone coronaria	Poppy flowered Anemone	Tuber/seed	October-November	35-40	March-April	Pink, red, purple, white, mauve
Begonia evansiana	Hardy Begonia	Bulbil/ tuber/seed	March-April	30-45	July-October	Vermillion, white and pink
Bergenia ligulata	Rock foil/ Elephant's Ear	Rhizome	November and February-March	25 -30	March-April	White with pinkish colour
Belamcanda chinensis	Blackberry Lily	Rhizome	October-November	90-120	August-September	Orange with crimson spots
Convallaria majalis	Lily of theValley	Creeping	October-November	15 -20	April-May	White
Crinum amoenum	Crinum/ Cape	Bulb/seed	March-April	30 -40	May-June	White
Crinum moori	Lily / Swamp Lily	Bulb/seed	March-April	60 -90	May-June	Soft rose
Crocus sativus	Spring Crocus	Corm	July-September	15-20	October-November	Lilac, purple
Crocus vernus	Autumn Crocus	Corm	October-November	15-20	April	Yellow, white, purple
Eremurus himalaicus	Fox-tail lily	Tuberous roots	October-November	125 - 150	April- May	Creamy-white
Fritillaria imperialis	Crown Imperial	Bulb/seed	August-November	60 -90	March-April	Brick-red, copper-red, orange

Scientific name	Common name	Mode of propagation	Planting time	Plant height	Flowering time	Bloom colour
Gerbera jamesonii	Transvaal Daisy	Division of clumps/ seed	April	30-40	May-October	Yellow, red, pink, white, bicolours
Galanthus nivalis	Snow drops	Bulb	November - October	10 -15	March-April	White
Hemerocallis species and *Hybrids*	Common day Lily	Division of clumps	October-November and March	80 -110	June-July	Yellow, orange
Hosta plantaginea	Platian lily	Rhizome	March	50-70	July-September	White
Paeonia lactiflora	Peony	Tuberous root	October-November	60 -80	May-June	White, red, pink
Platycodon grandiflorum	Balloon flower	Rhizome/ seed	March-April	45 -60	July-August	Blue
Nerine samiensis	Guernsey lily	Bulb	June-July	40 -50	September	Red, scarlet, yellow
Ornitho-galum umbellatum	Star-of-Bethelhem	Bulb	October-November	15 -20	March-April	White
Lilium regale	Regale lily	Bulb	October-November	50 70	June-July	White
Liatris spicata	Blazing star	Corm/seed	March-April	60 75	August-September	Purple
Muscari botryoides	Grape hyacinth	Bulb	October-November	15 -20	March-April	Sky-blue /white
Narcissus tazetta	Narcissus	Bulb	July-September	30-50	February-May	Yellow, ora-nge, white
Iris hollandica	Dutch iris	Bulb	October-November	40-50	April-May	Blue, white, purple, yellow
Kniphofia uvaria	Torch lily/ Red-Hot	Rhizome	March-April	80 -120	June-October	Coral-Red with yellow zome
Ranunculus asiaticus	Double-flowered Persian Butter cup	Tuber/ seed	November	30 -45	April-May	Yellow, white, red, purple, orange
Sternbergia lutea	Winter Daffodil	Bulb	September-October	15 -20	February-March	Yellow
Tritonia crocosmijlora	Montbretia	Corm	February-March	60-90	June-September	Red
Zantedeschia aethiopica	Calla lily	Rhizome	February-March	40 -60	May-June	White

DISEASES AND INSECT PESTS OF BULBOUS PLANTS

Bulbous flower crops are grown in commercial fields, greenhouses, landscape areas, home grounds as well as inside homes. Various diseases, insects and pests that cause variable losses in terms of quality as well as quantity both in field and storage, are attracted to bulbous plants just as to flowers. The climate under which plants are grown has great influence on the pest incidence and damage. The diseases caused by fungi, bacteria and viruses results in considerable damage, and loss of bulbs and flowers. Viral diseases are the most distressing for the bulbs of bulbous flower crops. If not checked in time, these can cause considerable economic loss. The diseases of important bulbous crops are given in Table 2. Different species of Aphids such as *Aphis gossypii, A. craccivora, Myzus persicae, M. gossypii, Macrosiphum solanifolii,* etc. (Asjes *et al.,* 1997), mites such as *Tetranychus urticae, T. telarius, T. bimaculatus, T. equatoirious, T. turkestani* etc. (Liu and Liu, 1998); thrips such as *Liothrips* species, *Thrips hawaiiensis, Taeniothrips simplex, frankliniella occidentalis* often invade most of the bulbous plants. The insect-pests of important bulbous flower crops are given in Table 3.

Table 2. Diseases of bulbous flower crops

Flower Crop	Disease	Causal Organism	Symptoms	Control
Gladiolus	Fusarium wilt	*Fusarium oxysporium* f. sp. *gladioli*	Curving, bending, arching, stunting, yellowing or drying of leaves associated with root and corm rots in field as well as in storage. Dark colour streaks on corm base.	Thiram, Captan sprays. Hot water treatment of corms and cormels contacting Carbendazim and captan
	Curvularia Blight	*Curvularia trifolii* f. sp. *gladioli*	Yellowing and browning of leaf tips Small to medium red spots which are dark brown in colour, rounded, elongated or irregularly shaped surrounded by a reddish brown ring, mostly shallow.	Spray 0.2% Dithane M-45 and miltox (Copper oxychloride +zineb) each at fortnightly intervals
	Bacterial Scab blight	*Xanthomonas gummisudans*	Rotting of leaf sheath and stem at soil level.	Hot water treatment of corms and cormels
	Neck rot	*Pseudomonas marginata*	Production of short, poor quality flower spikes and small corms.	

Flower Crop	Disease	Causal Organism	Symptoms	Control
	Botrytis blight	*Botrytis gladiolorum*	Small circular reddish brown sunken spots and results in spongy tissue during rot	Roughing of affected plants and removal of all infected debris Hot water treatment of corms and cormels Application of mancozeb
Lilium	Foot rot	*Phytophthora cactorum*	Infected parts become shrunken, plants wither and die, but bulbs are not damaged.	Proper drainage should be provided. Dig up and destroy infected plants Spray copper fungicide in early spring. 2% Bordeaux mixture
	Gray Mould	*Botrytis elliptica*	Presence of circular or oval and yellowish to reddish brown spots on the leaves. Central part is light grey in colour while the outer region is dark purple, shading into green healthy tissue.	Spray copper fungicide in early spring. 2% Bordeaux mixture
	Bulb rot	*Fusarium oxysporum* f. sp. *lilii*	Confined to the base of the scales, which are detached from the basal part of the bulbs Lower leaves grow from infected bulbs, become yellow or purple and dry prematurely. In severe infection, plants become stunted having extensively rooted scales.	Removal of infected plants. Drenching of soil with formalin. Captan application to soil is beneficial
	Bacterial Soft rot	*Pectobacterium carotovorum*	Soft and wet decay of bulbs	Discard infected bulbs and provide proper drainage. Avoid bulb injury from soil
	Viral Disease (Mosaic)	*Tulip breaking virus* *Cucumber mosaic virus* *Lily symptomless virus*	Dwarf and crippled plants Stunted and mottled plants	Use clean equipments

Flower Crop	Disease	Causal Organism	Symptoms	Control
	Rosette	Virus	Flat, rosette appearance of the plants Leaves are pale green or yellowish without mottling Downward curling of leaves	Use clean equipments Control aphids using systemic insecticides
Zantedeschia	Crown rot	*Pellicularia filamentosa* *Sclerotium delphinii*	Soft decay of the lower half of corm and feeding roots	Use of disease free corm for planting and soil sterilization
	Root rot	*Phytophthora richardiae*	Yellowing of leaves, browning and wilting Flowers become malformed and turn brown	Treatment of clean dormant rhizome in hot water formaldehyde or $HgCl_2$.
	Spotted wilt virus	*Lethum australiense*	Chlorotic to pale streaks and lesions between veins on the leaves	Use clean and healthy planting material and equipments.
	Leaf spot	*Phoma* spp., *Alternaria*	Small spots which become larger and irregular in decayed areas	Infected leaves should be removed & fungicidal spray should be used
	Powdery mildew	*Leveillula taurica*	Noticed on all the aerial parts of plants.	Spraying of wettable sulphur (0.2%) or karathane (0.025%).
Tulip	bulb rot	*Fusarium oxysporum* f.sp. tulipae	Large brown patches on outer fleshy scale of infected bulbs. Produces gum, which is a sticky liquid; it later becomes darker and produces sour smell. Field planted crops show early senescence and shoots become purple.	Dip the bulbs in systemic insecticides immediately after harvesting.
	Tulip fire	*Botrytis Tulipae*	Infected leaves do not open, patches of severe leaf spotting occurs which leads to loss of leaf area and finally flowers	Dusting the bulbs with sulphur powder and soaking the bulbs in systemic fungicide before planting.
Narcissus	Basal rot	*Fusarium oxysporum* f. sp. *narcissii*	Plants are stunted and foliage dies prematurely.	Do not plant bulbs that have white or pink fungus on them.

Flower Crop	Disease	Causal Organism	Symptoms	Control
			The few roots present are brown and the basal plate is decayed. White or pink fungal growth develops between the bulb scales.	Plant hot water treated bulbs.
	Leaf Scorch	*Stagonospora curtisii*	Dark brown oval lesions and spots starting from the neck of the bulb to the leaf tip at emergence. Burnt appearance. Premature death of foliage.	Destroy plant debris and infected flowers. Hot water treatment + zineb (75% W.P.) spraying Spray Bordeaux mixture when leaves are 5-10 cm high.
	Smoulder	*Sclerotinia narcissicola*	Yellow, malformed and stunted shoots, rotting of plants	Carbendazim sprays applied 14-20 days before and after picking control the disease effectively.
	Fire	*Botryotinia polyblastis*	Flowers are spotted, turn brown and die. Small, tan, oval spots are formed near leaf tips. Leaves have bright yellow streaking near tips.	Thiophante or chlorothalonil spray
Iris	Bacterial soft rot	*Erwinia carotovora*	Appear in newly planted as well as mature Iris plantings. Foul-smelling soft decay of the rhizome.	Plant healthy rhizomes and eliminate the Iris borers.
	Crown Rot	*Sclerotium rolfsii*	Produces small spherical sclerotia on the rotted leaf bases.	Remove and destroy affected rhizomes.
	Bacterial Leaf Blight	*Xanthomonas tardicrescens.*	Irregular spots that first appear near the margins on the leaf tips. Brown spots become larger and develop whitish or grayish centres. The bacterial infection follows the leaf veins down the leaves and the splotches may run together.	Remove blighted leaves during the season and destroy all foliage in the fall.

Flower Crop	Disease	Causal Organism	Symptoms	Control
Dahlia	Powdery mildew	*Erysiphie polygoni and Erysiphie cechoracearum*	White to grey powdery mass appears on upper surface of leaves.	Spray 0.1% Karathne Spray 0.2% Carbendezim
	Botrytis blight	*Botrytis cinerea*	During dull, cloudy weather, young shoots and flower buds are infected. Fading and browning of petals.	Spray of Zineb, Ferbam
	Smut	*Entyloma dahliae*	Presence of more or less circular spots, first yellowish green later turns to brownish, then dry out.	Spray of Z-78, Benlate etc.
Day lily	Leaf spot	*Cercospora hemorocallis Hetreosporium iridis*	Greenish yellow spots on leave, which gradually enlarge and turn orange brown.	
	Leaf streak	*Collecephalus hemerocalli*	Necrotic spotting and streaking of leaves.	
Amaryllis	Red leaf spot	*Stagonospora curtisii*	Spots on leaves, which ultimately turn yellow and fall off.	Spray 0.75% copper oxichloride. Warm water treatment of bulbs@ 43.5°C for 2 hrs.
	Soft rot	*Rhizopus stolonifer*	Spots on leaves, which ultimately turn yellow and fall off.	Spray 0.75% copper oxichloride. Warm water treatment of bulbs@ 43.5°C for 2 hrs.
	Bulb decay	*Sclerotium rolfsii*	Spots on leaves, which ultimately turn yellow and fall off.	Spray 0.75% copper oxichloride. Warm water treatment of bulbs@ 43.5°C for 2 hrs.
	Amaryllis mosaic	Cucumber mosaic virus	Large yellowish green patches appear at random over the leaf surfaces.	To control vector, 0.2% malathion spray at 15 days interval.
	Tomato spotted wilt virus		Numerous yellow spots on affected plants and they died ultimately.	Control of vector by spraying 0.2 % Rogor or malathion
Canna	Virus		Crinkled and distorted leaves and also show signs of speckling or streaking.	Burn all affected cannas.
	Botrytis blight		Brown or spotted flowers. Silver gray spores show up on the dead or	Roughing of affected plants and removal of all

Flower Crop	Disease	Causal Organism	Symptoms	Control
			decaying plant matter, if the plant is heavily infected.	infected debris. Application of mancozeb.
	Canna rust		Leaves and stems have orange spots. The orange spots are the fruiting bodies of the fungus rust.	Remove and discard affected foliage.

Table 3. Insects of flower crops

Flower Crop	Insect	Scientific Name	Damage	Control
Gladiolus	Seed Corn Maggot	*Hylemya cilicura*	Maggots feed on the seeds of Gladioli by entering the capsules in the field. Web the seeds in the storage.	Spray with 4% wettable chlordane powder. Dip the corms for one hour in thiram and insecticide solution for 24 hrs before planting.
	Aphids	*Aphis gossypii* *Myzus* sp.	Suck saps from young plants. Transmit viral diseases.	Spray metasystox, rogor, etc.
	Thrips	*Taeniothrips simplex*	Feed on flowers and floral sheeths. Cause silvering and whitish streaks.	Spray metasystox, rogor etc.
	Cut worms	*Agrotis segetum*	Feed on different parts of the plants and sever the plants at the base.	Spray metasystox, rogor etc.
	Mites	*Rhizoglyphus echinopus* *Tetranychus sp.*	Feed on leaves under warm & shady locations. White specks appear on leaves, which later turn into bronzing or silvering.	Spray miticides
	Nematodes	*Meloidogyne* sp. and *Trichodorus* sp.	Attack corms and roots.	Treat dormant corms in hot water (35°C) for 30 mintutes.
Lilium	Aphids	*Aphis gossypii* *Myzus* sp.	Suck saps from young plants. Transmit viral diseases.	Spray metasystox, rogor etc.
	Lily beetle	*Lilioceris lilii*	Feed on leaves, stems and seed pods.	Spray malathion.
	Thrips	*Liothrips vaneeckei*	Feed on bulbs. Produce rust coloured	Store bulbs at temperature below

Flower Crop	Insect	Scientific Name	Damage	Control
		Taeniothrips simplex	sunken areas at the base of outer scales. At later stages, these scales become soft and break off easily.	10°C. Dip the bulbs in hot water at 35°C for an hour.
	Bulb mite	*Rhizoglyphus echinopus*	Feed on bulbs and destroy the roots. Infest scales and leaves and stems become yellow.	Grow lilium in properly drained soil, Follow crop rotation. Store bulbs at 2°C Dip the bulbs in hot water at 35°C for an hour.
	Stem borer	*Papaipema nebris*	Bores in lily stems and feeds on pith.	Destruction of larvae by splitting the stalks open.
Narcissus	Large Narcissus fly	*Merodon equestris*	Maggots feed on bulbs and make tunnels in bulbs. Later on bulb rot because of fungal or bacterial infection.	Hot water treatment with 0.2-0.4% aldrin for 60 minutes. Application of temik granules at 40 kg a.i./Ha once in January.
	Bulb scale mites	*Steneotarsonemus laticeps*	Infested bulbs become undersized and soft and the shoot emerging from such bulbs produce initially bright green foliage, which turns later striped yellow.	Hot water treatment of bulbs
Dahlia	Aphids	*Aphis gossypii* *Myzus* sp.	Suck saps from young plants. Transmit viral diseases.	Spray metasystox, rogor etc.
	Thrips	*Taeniothrips simplex*	Feed on flowers and floral sheeths. Cause silvering and whitish streaks.	Spray metas-ystox, rogor etc.
	Mites	*Polyphagotarsonemus latus* *Tetranychus urticae*	Damage of foliage by feeding underside of leaves causing white speckling	Spray of meta-systox, wettable sulfur, kerathane, etc.
Tulip	Thrips	*Aeria tulipae*	Feed on flowers and floral sheeths. Cause silvering and whitish streaks.	Treatment of bulbs at the time of planting with chlorpyriphos.

Flower Crop	Insect	Scientific Name	Damage	Control
	Scale	*Hemiberiesia neodiffinis*	Infested bulbs become undersized and soft and the shoot emerging from such bulbs produce initially bright green foliage which turns later striped yellow.	Hot water treatment of bulbs.
Amaryllis	Mites	*Tetrany-nchus bimaculatus* *Rhizoglyphus echinopsus*	Suck sap from beneath the surface of leaves.	Hot water treat-ment of bulbs at 40°C for 60 minute.
Iris	Iris borer	*Macronocta ontusa*	Makes larval galleries in the rhizome and number of larvae found inside the galleries.	Application of dimethoate is effective against borer.

REFERENCES

1. Alice Aubrey 2001 *Zantedeschia aethiopica*, www.plantzafrica.com
2. Asjes CJ, Lilien KH, Borochov A, and Halevy AH 1997 Incidence and control of viruses in gladiolus in the Netherlands, *Acta Hort.* **430** 699-708
3. Basker D and Negbi M 1983 The uses of saffron, *Econ Bot.* **37** 228-236
4. Bhalla R and Sullhan B 2003 Tulips, in *Bulbous Ornamental* (eds) PK Rajeevan, KP Singh and PK Valsalakumari, 86-93 pages
5. De Hertogh AA and Nard M Le 1993 *The physiology of flower bulbs*, Elsevier, Amsterdam, 811 pages
6. Halevy AH 1990 Recent advances in control of flowering and growth habit of genotypes, *Acta Horticultrae* **266** 35-42
7. Janakiram T, Naveen Kumar P and Misra RL 2007 Gladiolus, in *Indian varieties of flower crops*, 35-41 pages
8. Jindal SL 1968 *Ornamental Bulbous Plants.* Indian Council of Agricultural Research, New Delhi, India, 190 pages
9. Liu TS and Liu TS 1998 Insects pests of gladiolus and lily plants,139-150. *The proceedings of the symposium on research and development of gladiolus, lilies and chrysanthemums*, Taichung, Taiwan, Special publication, Taichung District Agricultural Improvement Station
10. Misra RL, Kumar Ajith and Sharma Sanjeev Kumar 2003 Narcissus, in *Bulbous Ornamental* (eds) PK Rajeevan, KP Singh and PK Valsalakumari, 65-68 pages
11. Motial VS 1982 *The Canna* (1st edition), Kashmere gate, New Delhi: Cambridge Press, 13-17 pages

12. Panwar RS and Saxena NK 2001 Gladiolus cultivation, in *Commercial flower production* (eds) SS Sindhu and ML Choudhary, New Delhi, India: IARI, 35-40 pages

13. Raj Desh 2003 Alstroemeria, 5-8. *In* Rajeevan PK, Singh KP and Valsalakumari P K (ed), Bulbous ornamentals, New Delhi, India: IARI

14. Randhawa GS and Mukhopadhyay A 2004 *Floriculture in India.* New Delhi, India: Allied Publishers Private Limited

15. Rees AR 1984 Evaluation of the genotypic habit and its physiological advantages. *Herbertia* **45** 104-110

16. Smith FF 1963 Conrolling insects of flowers, *Agriculture Information Bulletin* No 237, U.S. department of Agriculture, 30 pages.

17. Vinayananda S 1986 Dahlia, 78-85. *In* Chadha KL and Choudhury B (ed) *Ornamental Horticulture in India*, New Delhi, India: Rekha printers.

LAWN

S. S. Sindhu

INTRODUCTION

Lawn is a green, carpeted patch of cultivated and tended grass that lends grace and dignity to a dwelling and provides varied avenues of enjoyment to the house holders; from aesthetics to partying to meditation. Naturally, its size varies in keeping with the dimensions of the house but should not be considered as an integral item in the overall scheme of the desired ambience. In this chapter, treatment of lawn is approached taking into consideration two aspects - the grass species, and the cultivation and maintenance requirements.

There are approximately 620 genera and 10,000 species of grasses worldwide. Grasses are extensively deployed throughout the world for their beauty and utility. Grasses are adapted to varied environments; from temperate to tropical to arid semi arid. They all belong to family *Gramineae.*

Fig. 5. Lawn

UTILITY OF LAWNS

(a) Utility of lawns: Grasses of lawns are pleasing to the eyes and soothing to the soul. They provide appropriate surface to play. They provide healthy surroundings and congenial environment for family interactions.

(b) Other uses of lawn: A lawn adds space to the house and surroundings, provides varied utilities including improving the quality of various ground features, offering a heathy environment, and adding value to the property.

SOIL PREPARATION AND STERILIZATION

Physical and Chemical Nature of Soil

Well drained fertile soil, with good water holding capacity, free from weeds, rocks and crocks in sub soil, is essential to develop an ideal lawn. The optimum pH of soil should be 5.5-7.0. Many warm season grasses grow better at pH 6-7.

Initial Planning

The steps to be taken in estabilishing a new lawn depend on the nature of site and the kind of lawn that is intended to be estabilished. This should be decided taking into account how the lawn is to be used and looked after later on. A clear vision is necessary so that the time and resourses invested produce the object desired.

Selection of Site and Shape

To develop a good lawn, soil should be preferably well drained, sandy loam and rich in organic matter. Sunny spaces are considered good for lawn. The shape of the lawn may be regular or irregular. However, rectangular shape is most popular. Graceful curves and slopes can add beauty to the lawn and the landscape.

Drainage

Grasses are shallow-rooted herbs and therefore, no deep drainage is necessary but stagnation of water in rooting zone is not acceptable. In clayey soil, some kind of drainage is necessary and may be achieved by laying drainage pipes or by putting a layer of broken pieces of bricks etc. 90 cm below the surface. Ordinary drainage work can be carried out in conjunction with grading or leveling. Generally 2-3 cm slope is necessary per 50 m towards the natural direction of drainage.

Digging

A rough estimate of surface levelling , by visualising the site, should be done prior to commencing the digging. If during rough levelling, a lot of shifting and filling of soil is visualised, the surface soil should first be taken out and kept separately so that it can be laid on the top after final levelling.

The digging operation should be done by the trenching method. A trench 60 cm deep and 45 cm wide should be dug for the entire length at one end of the site and the soil should be kept outside the lawn area. The rest of the area should be dug at the same depth in widths of 45 cm at a time, the earth removed from the freshly dug trench should go into the trench dug in the preceding operations. The earth first dug will go into the trench dug at the farthest end. At each stage clods should be broken and pulverized thoroughly. All stones, old masonry, grass roots etc. should be removed. Special care should be taken to remove the roots of motha grass or nut grass (*Cyperus rotundus*).

Digging is preferred during summer. After trenching, soil is left to dry in the scorching Sun for a period of 7-15 days to kill the weeds or insects. The soil should be subsequently turned 2-3 times at weekly intervals and clods should be broken. The roots of weeds should be removed.

Lawn with a gentle slope is ideal because it drains wellwhich is hardly noticeable to the eye). There are pros and cons to having undulations in a lawn though they are acceptable for large lawns provided the gradient is gentle and smooth enough to allow easy and efficient mowing.

Manuring

If the soil is originally fertile, it is better not to apply any organic manure, as organic manure contains weed seeds. But poor soil will need proper dressing with organic manure and well rotten cowdung manure mixed thoroughly at the rate of 5 kg per square metre and followed by rolling. Apply super phosphate or D.A.P. or bone meal in soil mixture, and mix well into the top layer of the soil by racking. Night soil manure, F.Y.M. or old stable manure is also used for this purpose. The manure is sieved finely and spread over the surface at the rate of 500 kg per 100 square metre of soil. The amount can be reduced depending on the soil fertility level. This is then worked up in the soil to a depth of 15-20 cm.

Soil fertility is much more than merely the right amount of mineral nutrients supplied by fertilizers. Aeration, moisture retention, organic content and pH are also important. Moreover, of the many mineral nutrients required by grass for satisfactory growth, only three commonly need supplementing - nitrogen, phosphate and potash.

Grading and Leveling

Any freshly dug up soil takes a long time to consolidate. The next step is to settle the soil thoroughly. In heavy rainfall areas, pouring rains do this well. In other areas, the prepared soil is watered heavily after providing bunds to prevent run off. The accumulated water in pools will show the area of depression, which should be brought to level by shifting soil from one place to another. The flooding should be repeated 2 or 3 times removing the sprouted weeds each time.

The final levelling is done with the help of levelling pegs, straight-edge and spirit level. A straight-edge is a straight wooden plank of 3 m length and 10 cm width. Strong split bamboo pegs of 30 cm length pointed at the bottom are paint marked 15 cm from the top. These pegs are driven in the ground up to the 15 cm mark, starting from the centre of the lawn, in parallel lines of 2.9 m and keeping the same distance between the pegs in a line. At a time 4-5 lines of pegs are put. Then the straight-edge is placed on two pegs of the central line, one edge of it resting on one peg and the other on the other peg of the same line. The spirit level is then placed on the centre of the straight edge and the position of the bubble is adjusted in the centre of its run by driving in or lifting the outer peg until the bubble is in the centre. Following the same method from one peg to other, as a first step, the level in the central line of pegs is corrected. Then one end of the straight-edge is placed on a corrected peg in the central line and the other end on the corresponding peg in the adjacent line and the level is corrected by lifting or driving the peg in the adjacent line. Proceeding in this way, all the pegs are fixed in the correct level. After this, earth is removed around the pegs on which the 15 cm mark has gone underground to make this visible. Then lines of strings are passed from peg to peg in a criss-cross way through these 15 cm markings and the soil surface is made parallel to the lines in all directions. The operation is time consuming and needs a lot of attention. The levelling work can be done by drawing slats of wood measuring 45 x 7 x 2.5 cm along the surface. The same procedure is followed until the whole area is levelled. The loose soil is then firmed up by using light rollers.

The soil is then lightly irrigated and the levels rechecked when the soil is sufficiently dried up. There may be slight surface settlings after this, which are corrected and pegs removed. It is always advisable to keep the level of the lawn 5 cm below the levels of paths and drives. The margins along the paths are raised by gradual slope of 15-20 cm to form a turf edge of 3-4 cm height. This method will help to keep the paths dry when the lawn is flooded with water.

It is not always necessary to have a perfectly levelled lawn. Lawns can be laid in undulated fashion also and such lawns look very beautiful. But, there should not be any depression because water will collect in depressions and kill the grass. Moreover, the slopes and mounds in a lawn should be gradual and artistic to simulate nature.

Thus, before planting, ground should be levelled to give a gentle slope for drainage of extra water. There should be provision of an outlet to drain off extra water. It is recommended to use pre-emergence weedicide Glyphosate (Round up) @ 2-2.5% to control nut grass (Cyprus rotundas) and other weeds. A weedicide should either be sprayed or mixed with flood irrigation a fortnight before planting of grass and the ground should be covered with polyethylene if possible.

SELECTION OF GRASS

Selection of grass is based on the climate where lawn is to be developed.

(a) For Warmer Climate

i. Bermuda grass (*Cynodon dactylon*): Bermuda grass, commonly referred as Doob grass, is widely used for planting lawn due to its fast growth, hardiness, tolerance to frost and smaller water requirements. It makes an excellent turf and responds well to frequent mowing. The appearance of stolon is very common in Doob grass if mowing is not done at weekly intervals especially during summer season. The recent trend is to use a strain of Bermuda grass known as Selection-1, which is very fine, soft and easy to maintain but, it does grow well only under sunny conditions. It forms high quality carpet grass.

ii. Japanese grass (*Zoysia Japonica*): This type of grass is highly suitable for smaller areas and home lawns. It grows slowly and hence frequent mowing is not required. It forms a high quality lawn but it is slow growing.

iii. Mexican grass (*Stipa tenuissima*): It is very fine grass with shining surface but it requires a lot of maintenance. It does not grow equally and uniformly. It grows in bunches and hence, mowing become difficult.

(b) For Cooler Climate

Certain grass strains of *Agrostis tenuis, Festuca rubra, Poa* species (also called Blue grass) are suitable for temperate climate. The common species *Poa pratensis*, which is also known as Kentucky blue grass comes up on its own. *Lolium perenne* (Ryegrass) is used under temperate climate. These species do well under cold climate, low sunshine and shaded place (along walls and underneath the trees). A detailed description of different cool season grass species is given below:

i. Creeping bentgrass (*Agrostis palustris*): It is extensively used in cool climates for golf course putting greens and tees, lawn bowling greens, grass tennis courts, and some home lawns. It produces a fine-textured, soft, very dense, carpet-like lawn. However, it must be carefully tended or it can quickly lose its attractiveness. It requires frequent watering, low mowing.and high nitrogen (1/2 g per square feet per month of active growth). It is susceptible to a wide range of diseases but some cultivars that are quite resistant are available and include 'Penncross', which is dense, disease resistant, and quick to recover from damage.

Fig. 6. Creeping lawn

ii. Kentucky Bluegrass (*Poa pratensis*): It is the most widely planted cool-season grass, especially in the northern latitudes. Blue-green in color, medium to fine textured and very cold hardy, it represents the standard for appearance against which other cool-season grasses are measured. It is widely used for lawns, athletic fields, golf fairways, and general-purpose turfs because of its beauty. Even so, it requires conscientious maintenance. It should be mowed to between 11/2 and 21/2 inches slightly higher during hot weather. It should be watered frequently and fertilized generously. Among the most useful cultivars is 'Adel phi', which is low growing, dark green and fine leafed; it is disease resistant, fairly tolerant of shade and heat, and attractive even under low maintenance.

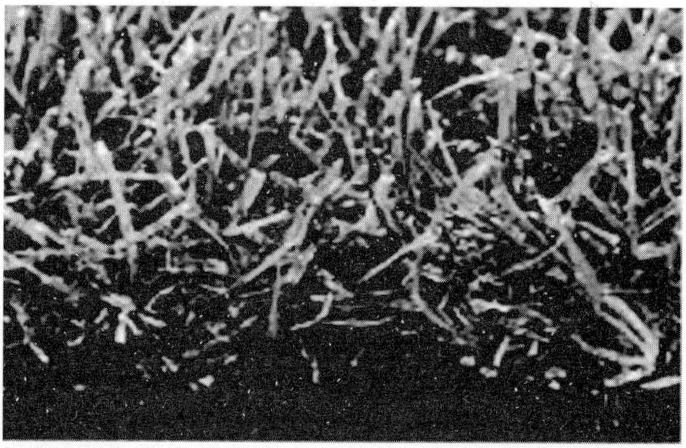

Fig. 7. Kentucky lawn

iii. Rough Bluegrass (*Poa trivialis*): It is a bright green, fine-textured, and shallow-rooted grass with boat-shaped tips to its blades. This relative of Kentucky bluegrass is noted for its high tolerance of moist soils and shade. The grass is soft, cold hardy and retains its colour over winter in mild climates. It makes a good component in shady-lawn mixtures, but in sunny areas it tends to crowd out other worthwhile grasses, particularly Kentucky bluegrass and Perennial ryegrass. Rough bluegrass is grown in the same locations as in Kentucky bluegrass. Popular cultivars are 'Colt', 'Laser', and 'Sabre'.

Fig. 8. Rough Bluegrass lawn

iv. Fine Fescuesb (*Festuca species and varieties*): These include Chewings, Hard, and Red fescue, all very fine leafed grasses that are extensively used in seed blends for both sunny and shady situations. They germinate rapidly (7-14 days) and establish themselves quickly. They are medium green in colour and spread by tillers or short creeping rhizomes. During extended hot dry periods, fine fescues may lose their colour rapidly.

Fig. 9. Fine Fescuesb lawn

v. Chewings fescue (*Festuca rubra commutate*): This is an aggressive, bunch-type fine fescue that can overtake other grasses - a bad quality if you want to preserve these but good if you want to crowd out weeds. Because of its high shade tolerance, it is sometimes used to overseed shady lawns, often in mixtures with Perennial ryegrass. It is well adapted to the sandy, acidic, often infertile soils that are found in these regions.

Fig. 10. Chewings Fescue lawn

vi. Hard fescue (*Festuca ovina* var. *duriuscula*): This is a fine-textured grass mostly found in the northern area at high elevations. Growing in clumps, it is slower to fill in than Chewings and Red fescue, but it needs minimal maintenance when mature. It is shade tolerant in well drained soils and is fairly drought resistant and salt tolerant. Highly resistant to diseases such as leaf spot and red thread, it is generally healthier than other fine fescues. It also stays greener over summer, even during extended dry periods. Its wearability is fair, though the clumps do recover slowly from damage.

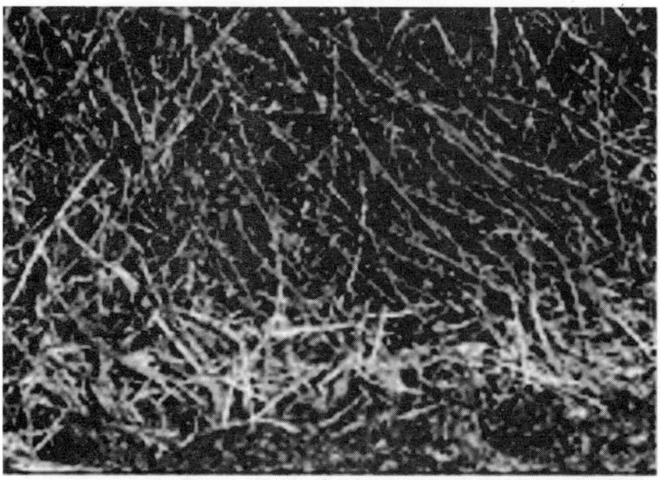

Fig. 11. Hard Fescue lawn

vii. Red fescue (*Festuca rubra*): It is also called 'Creeping Red fescue' and is a frequent component of bluegrass mixtures. A fine-textured grass with narrow, dark green blades, it blends well and does what some bluegrasses cannot; it grows well in both shade and drought. Red fescue is preferable over Chewings fescue in a seed mixture because it is more heat tolerant and less likely to form thatch. It is best adapted where summers are cool. It grows well on banks and slopes, and creates an especially lush effect when left unmoved.

Fig. 12. Red Fescue lawn

viii. Tall fescue (*Festuca arundiacea*): It is a dense clumping of grass that forms a coarse turf able to grow in Sun or shade. This species stays green all year in mild-winter climates. It is a good general grass for home lawns as well as for playing fields and commercial grounds. Its disadvantages are its coarse texture and clumping style of growth. It does best in areas of mild winters and warm summers.

Fig. 13. Tail Fescue lawn

ix. Annual ryegrass, Italian ryegrass (*Lolium multiflorum*): Cool-season annual suits in cooler climates, a short-lived perennial bunch-type grass. It forms a medium to coarse-textured lawn with moderate wear resistance. In temperate climates, it is sometimes used as a temporary lawn in late spring. Best grown in full sun, this grass requires a moderate to large amount of water and is not drought tolerant. It also has poor tolerance of heat and cold.

Fig. 14. Annual ryegrass

x. Perennial ryegrass (*Lolium perenne*): Deep green glossy leaf blades, a shallow root system, and a texture that is finer than Annual ryegrass characterize this grass. Exhibiting the best wear tolerance of any cool-season grass, it is often used as a tough play lawn. However, its intolerance of extreme heat, cold, and drought makes it best adapted to regions with mild winters and cool, moist summers. It is sometimes used instead of Annual ryegrass to oversee dormant lawns of warm-season Bermuda grass during winter. But unlike Annual ryegrass, it tends to persist during the transition from cool to warm weather. This is a disadvantage if you want the Bermuda grass to predominate again when summer returns. Perennial ryegrass likes full Sun but can tolerate some shade. Its non-creeping, bunchy-type growth forms a uniform lawn if the grass is properly established and maintained.

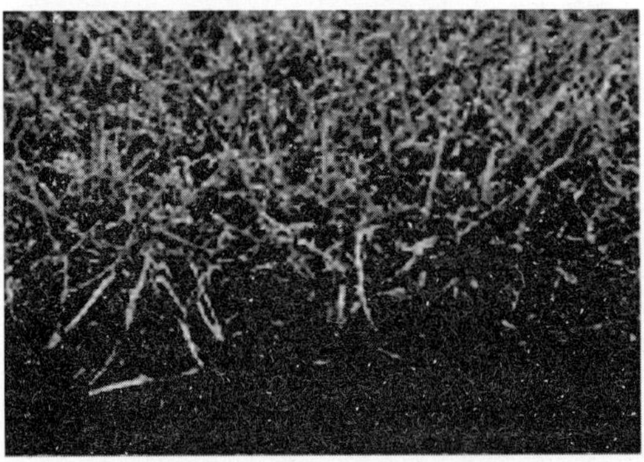

Fig. 15. Perinnial ryegrass

METHODS OF PLANTING

(a) Seed Method

2.5 gm of seed per square metre is optimum, which is generally mixed with garden soil or sand for even distribution and spread in the area. It is followed by slight raking of the soil for covering the seeds. Sprinkle water immediately after sowing and frequently thereafter till seeds are fully germinated.

(b) Dibbling Method

This is most widely used practice in India. Grass pieces along with roots are taken from weed free areas and dibbled at a distance of 10-15 cm both ways followed by immediate irrigation.

(c) Turfing Method

This is the quickest method of making a lawn within a day. The intact pieces of uniform thickness of turf with soil are taken from the place where grass is compact and free from weeds. These pieces of carpet are evenly spread on the ground and covered with fine soil followed by watering.

TOP DRESSING

In lawn management, top dressing means the application of bulky material like sandy compost, spread as evenly as possible at 1.5–3.0 kg per square metre by hand or with a shovel and then worked into the sole of the turf using suitable equipments such as a drag brush, drag mat, lute or even simply the back of a wooden rake.

Top Dressing Material

The ideal material for top dressing is a special form of sandy compost produced by making a compost heap with alternating layers of sandy soil and rotted farmyard manure. It is allowed to stand for a year or so before breaking down and mixed with suitable sand to produce material with the right sandy texture. Obviously, this procedure is not very practicable for the lawn owner who, like many professionals, make up for an acceptable substitute. This might well take the form of a mixture of:

- 6 parts suitable sand
- 3 parts sandy loam soil
- 1 part granulated peat

Good, clean (weed-free) sandy loam should be obtained for the mixture but other types of soil can be used with suitable adjustment of the proportion of sand added i.e. less sand for a sandy soil, more sand for a heavy soil. Really heavy or clay soil should be avoided since it is almost impossible to produce a good uniform friable mixture from them. Organic matter for the mixture is conveniently provided by crumbly granulated sphagnum or sedge peat, but other organic materials such as leaf mould may also be used, again possibly with an adjustment of the proportions. For example, the leaf mould may be of a decidedly soily nature so that the proportion of sand may need to be increased.

Top dressing with sand alone or peat alone is not recommended since it can ultimately result in the formation of layers as in a sandwich cake and this layering is detrimental to root development, drainage and drought resistance. By regular use of the right kind of top dressing mixture, it is possible over a period of time, to reach ideal growth conditions irrespective of the original state of the soil, and grow the turf in what can be called as perfect top soil. The most convenient time to apply top dressing is early autumn while some growth persists. However, top dressing may be carried out at any convenient time in the growing season when the grass can push its way through and anchor down the compost. It is unwise to top dress in the winter months when there is little or no growth.

Maintenance

Rolling, Mowing and Sweeping

The objective of rolling is to help the grass anchor itself securely and also to keep the surface levelled. Rolling should be avoided when the soil is wet. There is difference of opinion regarding the rolling of Doob lawns. It is argued that Doob grass has a growth pattern completely different from the English lawn grass. The Doob grass needs rolling at regular intervals while the English lawn grass does not. Moreover, rolling does not help much in levelling. Instead it leads to caking or hardening of the soil surface. People with the opposite view suggest very light rolling at the time of levelling of the ground and once after the lawn has just established to press down the small stones or pebbles, which otherwise interfere with the mower blades. This view is justified especially for heavy soils. In light sandy soils rolling after each weeding will be helpful to keep the surface levelled. Indiscriminate use of a roller is definitely bad.

Mowing is another important operation for the proper maintenance of a good lawn. Newly planted lawns are not generally mowed till they get established well. The first thing is to obtain a good machine, which will cut evenly at a correct height. The frequency of mowing is determined by the amount of growth and will vary from season to season. But grass should not be allowed to grow more than 5-6 cm in length during any season.

Sweeping the lawn thoroughly after each mowing is essential to clean the cut grasses, which might have fallen from the mower box. Sweeping should also be done every morning to clean the fallen leaves and other debris. Sweeping should be repeated two or three times in a day during the season when the deciduous trees shed their leaves.

Lawn Edge Trimming

After mowing, appearance of the lawn is greatly improved if the grass growing over the edges is trimmed. A pair of garden shears may be satisfactory for a very small lawn but for a larger lawn something less back-breaking is well worthwhile.

Weeding

One of the main aspects of maintenance is the control of weeds. Weeds are common in both new and old lawn. Therefore, as soon as a lawn is established weeding should start and continue at regular intervals. Motha, also called as nut grass (*Cyperus rotundus),* is very tenacious and common weed of lawn, and it is difficult to eradicate. Motha should be removed with the roots with a long narrow bladed (1-1.5 cm) *Khurpa.* Hand weeding is possible in small areas but for large lawns application of weedicides becomes necessary. For dicot weeds, 0.5% solution of 2, 4-D is recommended. Alternatively higher concentration of Glyphosate (Round up) or Stomp weedicides may also be used. But, once the lawn is well established, regular mowing keeps the weeds in check. All weeds should be removed along with the roots and should never be allowed to seed.

Irrigation

Lawn requires plenty of water. During winter months, watering should be done at an interval of 7 days whereas during summer season, it should be done on alternate days. Labour as well as water can be saved to a considerable extent if sprinkler irrigation is used.

Scrapping and Raking of Lawn

Continuous rolling, treading and mowing may result in the formation of a hard crust and the lower part of the lawn may get matted and woody. For such lawns, the grass should be scraped at the ground level with the help of a khurpi or with a diesel lawn mower by setting it at zero level during the months of April and May. It should be followed by application of a mixture of sand, soil, bone meal and fertilizers in equal ratio, and watering. Scraping should be followed by raking to break the crust. Where the condition of the lawn is good, hard and thorough raking should be done both ways to loosen the old runners and to aerate the soil.

Fig. 16. A view of the lawn

Top Dressing with Compost and Fertilization

After scraping or raking, compost consisting of good garden soil, coarse sand and leaf-mould in the proportion of 1:2:1 (in sandy soil the proportion of sand should be reduced or eliminated altogether) should be spread over the lawn to a depth of 3-5 cm, which roughly works out to be 100 kg per 100 square metre area. Bonemeal should also be applied at the rate of 1 kg per 10 square metre area. The same compost should be used as top dressing again during September or October. From October to April, ammonium sulphate should be applied once every month at the rate of 1 kg per 50 square metre area followed by watering. The authors have found that the application of fermented town compost in liquid form is very good. This should be prepared by fermenting 20 kg of compost in 100 litre of water for a few days. During fermentation ammonium sulphate and superphosphate at the rate of 1 kg and 2 kg should be added to the mixture. The concentrated mixture should then be strained through gunny cloth and diluted to tea colour and added to the lawn with water cans. After the application, the lawn should be soaked with water. This can be applied twice a year (October and May-June). Raw cowdung can be fermented and used in the same way. Application of 500-600 g per square metre of a mixture (2 parts of calcium ammonium nitrate:1 part of super phosphate:1 part of potassium sulphate) during rainy season and in February–March, gives a good growth and green colour to grass. Spraying of urea at the rate of 0.2% has also been found useful. Ammonium sulphate 1 kg per 100 square metre every month from October to February gives lush green lawn with a thick growth like a carpet.

TOP DRESSING METHODS AND MATERIAL

In lawn management, top dressing means the application of bulky material such as sandy compost, spread evenly at 1.5-3.0 kg per square metre by hand or with a shovel and then worked into the sole of the turf by suitable equipment such as a drag brush, drag mat, lute or even simply the back of a wooden rake. The ideal material for top dressing is a special form of sandy compost produced by making a compost heap with alternating layers of sandy soil and rotted farmyard manure and allowing it to stand for a year. This compost is mixed with sand to produce material with the right sandy texture. A mixture of:

- 6 parts suitable sand
- 3 parts sandy loam soil
- 1 part granulated peat

By regular use of the right kind of top dressing mixture, it is possible over a period of time, to reach near ideal growth conditions irrespective of the original state of the soil. Top dressing may be carried out at any convenient time in the growing season when the grass can push its way through and anchor down the compost. It is unwise to top dress in the winter months when there is little or no growth.

Fertilizers

Few lawns need more than one dressing a year of a balanced fertilizer mixture. Three main mineral nutrients, nitrogen, phosphate and potash in suitable proportions in spring. For those who wish to make their own mixture, the following formula suits the average conditions:

	Per 100 m^2
Sulphate of ammonia	2.0 kg
Fine hoof and horn meal	0.9 kg
Dried blood	0.9 kg
Powdered superphosphate	2.0 kg
Fine bone meal	0.9 kg
Sulphate of potash	0.9 kg

Liquid Fertilizers

In recent years, some fertilizers are applied in solution form but these have limited effect, partly because of the requirement of solubility and partly because of the difficulty of ensuring that every bit of turf gets the same amount.

PLANT PROTECTION

Diseases

a. Fusarium Patch Disease

Fusarium patch disease is caused by the fungus *Fusarium nivale*. It occurs most frequently during mild, moist weather at almost any time of the year but especially in the spring and autumn, particularly the latter. The first signs are usually small irregularly shaped yellow patches (often overlooked), which under conducive conditions grow larger and darken in colour from reddish or bright brown (when the disease is most active) to dark brown when the grass looks rotten and is probably dead. When the disease is rampant, faintly pinkish white mycelium may be seen round the perimeter of the light brown patches. The disease needs surface moisture, so switching or brushing the turf to remove water droplets from the grass-leaves during disease-encouraging weather helps. Excessive amount of top dressing, whether over the whole lawn or only parts of it, also encourage disease attack by smothering the grass and holding moisture on it. If the disease has appeared and weather conditions are favourable for its spread, early application of fungicide is necessary. Various proprietary lawn fungicides are available for use against fusarium disease. Recommendations given by the manufacturers should be strictly followed when using fungicides and safety precautions should be observed regarding toxicity hazards.

b. Corticium Disease

Corticium disease (caused by *Corticium fuciforme*) appears during summer as brown patches of grass. On close observation, small pink or red branched needles of the fungus may be found on the grass blades and sheaths. In serious cases a suitable turf fungicide should be applied with due care.

c. Surface Basidiomycetes

A fungal trouble, that has become more prevalent in recent years and not yet acquired a common name, is caused by superficial Basidiomycetes. Theseattack surface fibrous material in which they can be seen as white mycelium. Several fungi are involved in its causation, one being *Cristella confines and* others are yet unidentified. They cause depressions in the surface. The fungi seems difficult to control but application of organic fungicides helps.

d. Fairy Rings

Fairy rings are caused by soil-inhabiting fungi belonging to the Basidiomycete family. Fairy rings start from a central point and spread out in ever-widening circles. The mycelium of the fungus ramifies in the soil and, in the worst case, maks it more or less waterproof. There are several types of fairy rings. The worst one (so called Grade I fairy ring) is caused by the fungus *Marasmius oreades*. This usually appears as two dark green rings with a brown or bare zone in between the rings. They may be quite small (less than a dinner plate) or very large indeed and stretching over whole field. The soil, impregnated to a depth of 100-300 mm (4-12 in) or more with white mycelium, smells strongly of mushrooms and the fruiting bodies, which appear on the surface. These are, in fact, tan-coloured toad-stools. The fungus is susceptible to a number of fungicides but in practice it is very difficult to eliminate, possibly because the fungus gains protection from the water-proofing effect it has on the soil, which prevents fungicides from penetrating sufficiently. These fairy rings are so difficult to get rid of that most people decide to tolerate them. The alternative is to dig out and take away all the affected turf and earth to a width of 300 mm (1 ft) clear of the green rings inside and outside the rings and to a depth of 300 mm (1 ft). Then preferably to sterilize the bottom and sides of the hole with formalin before replacing the excavated material with clean soil and new turf. Before embarking on such a vast undertaking, it is prudent to consult an expert.

Grade II fairy rings show dark green ribbons or rings of grass on which puff balls or mushrooms may occasionally be found. One of the several different types of fungi may be responsible for this and though they may dry out the soil, there is no brown or dead area because the dense white fungal mycelium does not go as deep into the soil as it does with Grade I rings. Once again people usually tolerate these, since their effect is negligible. In some cases they can be controlled by application (after spiking) of organic fungicide. At least, this prevents the production of fruiting bodies.

Grade III fairy rings are also produced by many types of fungi. There is no apparent effect on the grass but rings or groups of toad-stools may appear. Control is probably unnecessary since the toad-stools can be swept off or mowed off but local application of organic mercury fungicide can stop the trouble.

Control: Drenching of soil with Bordeaux mixture (4:4:50) or Bavistin powder or any other suitable fungicide.

Pests

a. Birds

Starlings, rooks, jackdaws and other birds pay attention to turf mainly when there are grubs or insects in it. They may therefore, serve as a warning that action is needed to deal with the infestation and, once the cause of their interest is removed, they soon go away.

b. Animals

Field mice

Occasionally small holes made by field mice can be found. They can be controlled by rapping or gassing. Sometimes filling the holes and making life uncomfortable for the little animals succeeds in driving them away.

c. Moles

Lawns in urban areas are seldom disturbed by moles. When they appear, effective control can be achieved through gassing the creatures with motor car exhaust fumes or through the use of special dusts, which release toxic gases when moistened. These are blown into the runs by a special kind of blower or gas pump obtainable from horticultural suppliers.

d. Rats

There are special baits available with the chemists, which can be effective although they must be used with great care. In some cases gassing for moles proves a better approach. Trapping is seldom practicable.

e. Earthworms

The most common natural pest is the earthworm. There are many species of earthworms but only a few of them produce casts of earth on the surface of turf. If a wormy patch appears on a lawn it may be wise to apply a suitable wormkiller at an early stage before the worms get multiplied and spread throughout the lawn. Worms are most active in warm, moist weather during late spring and early autumn. Therefore, worm-control treatments should be given in spring or autumn (preferably the latter) when the worms are near the surface and likely to encounter the control material before it is decomposed or dispersed. A product of vegetable origin, Derris powder, is fairly safe (except to fish) when applied as recommended for control. The rate of application for these products depends on the strength of the material obtained and it is usually best to water-in. Most of the worms are killed below ground but some come to the surface and need to be swept up. The control typically lasts for about a year. Lead arsenate powder for which watering-in is optional, can also be applied.

f. Ants

Small heaps of fine soil are deposited on the surface of the lawn mainly in summer. This is caused by Ants (*Lasius flavus),* usually yellow brown, which carry soil particles up to the surface and make an underground passage.

Control: Brush them away when soil is dry, water the nest, and dust the area with alrdin or thimate or any insecticide in powder form.

Other Problems

a. Pale or yellow lawn

Yellowing of lawn may be due to several reasons and hence proper care must be taken in diagnosing the problem. This may be due to water logging or poor drainage in heavy rainfall areas or also due to lack of water. Deficiency of nitrogen is also responsible for yellowing of the grass, which can be prevented by frequent application of nitrogenous fertilizers.

b. Frost-injury

In North India, the grass is injured due to frost. This can be avoided to a great extent if the grass is sprayed with water every evening and in the early morning after frost.

Common Tools Required for Maintenance of Lawn

a. Lawn mower: Manual and mechanical available in different sizes in the market

b. Roller: Wooden or metallic

c. Coring machine

d. Garden roller barrow

e. Sprinkler

f. Hose pipe

g. Hedge shear

h. Secateure

i. Khurpi

j. Spade

k. Water can

l. Drag mat and brush

m. Garden roller barrow

n. Lawn sprinkler

CONCLUSION

In a lawn, the maintenance is demanding but fairly simple. Regular attention is necessary; particularly for mowing which must be carried out as frequently as there is growth to justify it. In general, however, a lawn owner does not need vast technical knowledge which some seem to consider necessary.

Indeed, there are probably as many good lawns spoiled by their owners trying too hard as there are lawns spoiled by neglect. Once a lawn has been established to an acceptable standard, it is often possible to maintain it almost indefinitely by very little attention other than regular mowing.

FREQUENTLY ASKED QUESTIONS AND MODEL ANSWERS

What is the utility of a lawn?

Ans. The lawn is integral part of any garden. It has many uses in our daily life. The following are the different utilities of the lawn:

(a) Aesthetic utility: Grasses are pleasing to the eyes and soothing effect to the soul.

Sports and activity: Grasses provide appropriate surface for playing and sports. One can enjoy the life more, being with natural greens and open spaces far from crunchy city spaces.

Being with family: Lawn provides ultimate platform for family gatherings and enjoyment.

Body enrichment: Being with nature means living in oxygen-enriched climate where one can boast Aerobics and Yoga, and simply encourage time spent outside.

(b) Adds Space: Lawn adds space to house and surroundings, which is being normally utilized for various purpose such as sitting, holding get-togethers and other family functions.

(c) Lawn fetches all utilities what grasses possess such as:

- Improve water percolation and water absorption capacity of soil.
- Slow down soil erosion and rain-water runoff. Thus, improves physical and chemical properties of soil to sustain life.
- Uplift the beauty of the property and add value to the property.

List common grasses for warmer and cooler climate.

Ans. (a) For Warmer Climate:

i. Bermuda grass (*Cynodon dactylon*)

It is commonly referred to as Doob grass andis widely used to plant lawn due to its fast growth, hardiness, tolerance to frost, and less water requirements. It makes an excellent turf and respondes well to frequent mowing. The appearance of stolon is very common feature in Doob grass if moving is not done at weekly intervals especially during summer season. The recent trend is to use a strain of Bermuda grass known as Selection-1, which is very fine, soft and easy to maintain but it does grow well under sunny conditions. It froms high quality carpet grass.

ii. Japanese grass (*Zoysia Japonica*)

This type of grass is highly suitable for smaller areas and home lawns. Its growth is slow and hence frequent mowing is not required. It forms a high quality lawn but it is slow growing turf.

iii. Mexican grass

Mexican grass is very fine grass with shining surface but it requires a lot of maintenance. This grass does not grow equally and uniformly. It grows in bunches and hence, mowing become difficult.

(b) For Cooler Climate

Certain grass strains of *Agrostis tenuis, Festuca rubra, Poa* species (also called bluegrass) are suitable for temperate areas. The common species *Poa pratensis* (also known as Kentucky bluegrass grown on its own) and *Lolium perenne* (Ryegrass) are used very well under temperate climate. These species do well under cold climate, low sunshine, shaded places, along the walls, and underneath the trees. It remains lush green till the temperature rises above 30-35°C.

What should be rate of seed per acre?

Ans. 10 kg per Acre (2.5 g per square metre)

How to develop a lawn? Describe different methods of developing a lawn.

Ans. There are three methods to develop a lawn.

1. Seed: 2.5 gm of seed per square metre is optimum, which is generally mixed with garden soil or sand for even distribution and spread in the area. It should be followed by proper raking of the soil for covering the seeds. Water should be sprinkled immediately after sowing and continued till seeds get fully germinated.

2. Dibbling: This is most widely used practice in India. Grass pieces along with roots are taken from weed-free areas and dibbled at a distance of 10-15 cm both ways, followed by immediate irrigation.

3. Turfing: This is the quickest method of making a lawn within a day. The intact pieces of uniform thickness of turf with soil are taken from the place where grass is compact and free from weeds. These pieces of carpet are spread on the ground evenly and covered with fine soil followed by watering.

What is mowing and rolling? Why it is done in a good lawn?

Ans. Newly planted lawn is not mowed till it gets established well. In the beginning, the grass should be trimmed with the help of scythe. Light rolling should be done on dry ground to suppress the upright growth and to anchor the grass firmly in the soil and to maintain the level of the ground. Interval of mowing of an established lawn depends on the season. The height of the shears of lawn mowers should not be more than 5-6 cm. In rainy and winter season, mowing is done at an interval of 7-10 days, whereas in spring, the interval is of 15 days, and in summer, it is of a month.

What is the need of scrapping a lawn?

Ans. It is done to renovate the old lawn when it becomes old and grasses become compact. After 3-4 years during summer , lawn should be completely scraped with the help of khurpa and raking should be done both ways. Before the start of the rains, top dressing mixture consisting of garden soil, sand and sieved leaf mould (1:2:1) should be applied to cover the upper 3-5 cm surface. It helps in improving the renovation of stolon or rhizomes, fertility level and levelling of the ground. Bonemeal @ 1 kg per 100 square metre should be applied with the onset of rains or irrigate the lawn to make a luxuriantly healthy turf.

How to identify the fairy rings disease?

Ans. The disease symptoms are characterized by the appearance of circular rings of thin colour or dead grass. Sometimes ants, termites and soil borne pests also cause this problem. It can be controlled by drenching the soil with Bordeaux mixture (4:4:50) or Bavistin powder or any other suitable fungicide.

List the most common garden tools for lawn maintenance?

Ans. Lawn mower, Roller, Coring machine, Spring-tine wire rake, Drag mat and brush, Garden roller barrow, Lawn sprinkler, Half-moon edging iron, Hose pipe, Hedge shear, Tree cutter, Brush cutter, Wheel barrow, Khurpi, Phawara etc.

SUGGESTED READING

1. Arora JS 2004 Introductory Ornamental Horticulture, Kalyani Publishers
2. Bose TK and Yadav LP (ed.) 1989 *Commercial Floriculture*, Calcutta: Naya Prokash
3. Chadha KL and Choudhury B 1986 *Ornamental Horticulture in India*, New Delhi: ICAR
4. Crockett JU 1971 *Lawns and ground covers*, Alexandria, Virginia: The Life Books Publishers
5. Escritt JR 1979 *Lawns,* Hodder & Stoughton: Teach yourself books Publishers
6. Kher MA 1975 *Chrysanthemum,* New Delhi: Dutta Publishers
7. Larson RA 1980 *Introduction to Floriculture*, New York: Academic Press
8. Pal BP 2005 *The Rose in India*, New Delhi: ICAR
9. Randhawa MS 1976 *Gardening through ages*, New Delhi: The Macmillan Co. of India Ltd.
10. Sindhu SS and Choudhary ML 2001 *Commercial Flower Production*, New Delhi: Division of Floriculture & Landscaping, IARI
11. Swarup Vishnu 1967 *Garden flowers*, New Delhi: National Book Trust
12. Swarup Vishnu 1997 *Ornamental Horticulture*, New Delhi: The McMillan Co. of India Ltd.
13. Trivedi P 1987 *Home Gardening*, New Delhi: ICAR.

CACTI AND SUCCULENTS

Y. C. Gupta, Bhavya Bhargava and Priyanka Sharma

INTRODUCTION

Cacti and succulents are fascinating for their whimsical plant form, strange and ever changing shapes, remarkable range of sizes, attractive spines and hair,for their delicate flowers, which are a miracle of beauty, exquisite colours, fragile loveliness, design and delicious fragrance. Cacti and succulents are now the fashion, a collectors's item, both for scientific interest and ornamental value. They are neither difficult nor demanding in their requirements and one can grow and study a large number of cacti even in a limited space.

The cactus family is native to the Americas, where their range extends from Patagonia to the Canadian-American border, but they are densest and most diverse in the regions of Northern Mexico and the southern tropics of Argentina and Bolivia. Many other species have naturalized outside the Americas after being introduced by people, especially in Australia, Hawaii and the Mediterranean region. Cacti inhabit diverse regions, from coastal plains to high mountain areas, and from the sub-tropics to the deserts.

CLASSIFICATION

Takhtajan (1966) gave the classification of cacti and succulents, which is summarised below and which classifies the orders and families constituting cacti and succulents.

Kingdom	:	Plantae
Phylum	:	Tracheophyta
Division	:	Anthophyta and Magnoliophyta
Class	:	Magnoliopsida and Dicotyledonous
Subclass	:	Caryophyllidae
Order	:	Caryophyllales
Family	:	Cactaceae (cacti)

Genus : About 100 genera and 1000-2000 species divided into three tribes

Tribe : Pereskieae, Opunitieae and Cactoidea

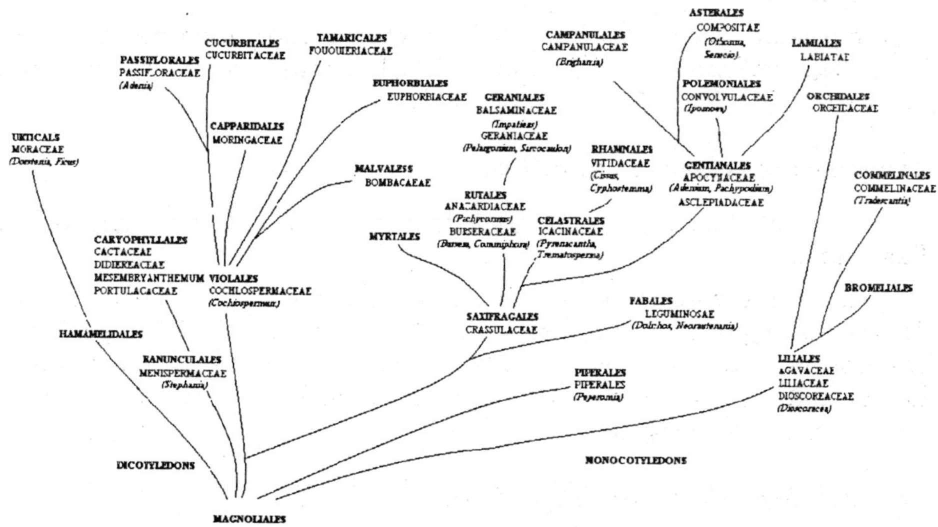

Fig. 17. Classification of Cacti and Succelents

- **Pereskieae**

Genera: Pereskia and Maihuenia

They have woody stems, persistent or semideciduous leaves, spiny and woolly areoles at the leaf axils, pendulous flowers and fleshy fruits.

- **Opunitieae**

Genera: Opuntia (prickly pear cactus), Pereskiopsis, Pterocactus, Quiabentia, Tacinga

These are fleshy, branched plants with flat or cylindrical leaves, seedy berry fruits and spines that are usually straight. Their areoles may or may not have spines but always have fine hair (glochids).

- **Cactoidea**

Genera: About 100 genera (most cacti are in this tribe)

The family Cactaceae comprises about 150 genera and 2,000 species They have developed a complex organ, known as the 'Areole', which is spirally arranged over the stem. This is a characteristic organ common to all Cactaceae and not found in any other plant. Areoles are the vital, growing area of cacti from which branches, leaves if any, spines, wool, flowers, new joints etc. arise. Stems are fleshy, with various shapes and sizes. They generally bear shallow and fleshy roots but occasionally, well developed tap-roots or tuberous are seen.

Important genera of the family Cactaceae are described below in brief. The species mentioned under each genus are interesting and attractive.

Acanthocereus Britton and Rose

8 species, South East America to Venezuela, Colombia and North East Brazil.

The Greek word 'Akanthos' means a thorn and 'cereus' means cactus. This is trailing or climbing cactus. The plants are usually 3-5 angled, and areoles are with short wool and hard spines. Flowers are long tubed, funnel-shaped, large, white and nocturnal. Fruits are spiny. *Acanthocereus pentagonus.*

Acanthorhipsalis Britton and Rose

Five species, South-east Peru and Argentina.

Plants are stocky, bushy and erect with mostly three cornered waxy green joints. They possess spines in the areoles. Flowers are golden orange in colour.

Acanthorhipsalis micrantha.

Aporocactus Lem.

Six species, Mexico and America.

The name is derived from the Greek, meaning 'impenetrable cactus' or commonly known as the 'Rat-tail Cactus'. They have slender, trailing stems which hang down gracefully. The stems may be 1.5 to 2.5 cm in diameter with 10-12 ribs. Flowers are funnel-shaped and crimson coloured. *Aporocactus conzatti*, *Aporocactus flagelliormis*, *Aporocactus flagriformis*, *Aporocactus leptophis*, *Aporocactus mallisonii*, *Aporocactus martianus.*

Arequipa Britton and Rose

Three species, Peru and North Chile.

They are globular to cylindrical cacti, up to 60 cm long, coated with long, coloured spines and wool. Flowers are funnel-shaped with a scaly tube and ovary.

Arequipa leucotricha.

Ariocarpus Schedw.

Five species, Mexico.

The name is derived from the Greek word 'aria' meaning the whitebeam and '*karpos*' a fruit. The fruit of these cacti resembles those of whitebeam, a kind of mountain ash. They are spineless, round, flat-topped with a taproot of low growth. A mature specimen is about 15 cm across and greyish in colour. The tubercles have the appearance of thick, triangular, fleshy leaves.

Ariocarpus retusus, Ariocarpus fissuratus.

Astrophytum Lemaire

Six species, South America, Mexico.

The name is derived from the Greek word '*aster*' meaning a star and '*phytos* a plant, in allusion to the flattered form of the cacti. The stems of these plants are globular, with pronounced ribs and covered with star-like scales. Spines are weak or wanting. Flowers are large.

Astrophytum asterias, Astrophytum capricorne, Astrophytum myriostigma, Astrophytum ornatum.

Aztekium Bodekar

One species, Mexico.

This is a globular plant with 8-10 ribs and 1-4 spikes. The flowers are white with a reddish tinge.

Aztekium ritteri.

Bortschella Britton and Rose

One species, Lower California.

These are oval-shaped, greyish-violet cacti, up to 10 cm tall. The top portion of the plant is woolly. Spines are white coloured, awl-shaped and tipped bucks. Flowers are purple or lavender in colour.

Bortschella schumanii.

Borzicactus Riccobons

17 species, Andes.

Plants are erect but a few are spreading type with many ribs which are cloud covered with areoles and spines. Flowers are small and round.

Borzicactus icosagonus, Borzicactus samaipatanus, Borzicactus sextonianus.

Cactus L. emend. Britton and Rose

40 species, South America and West Indies.

Name applied by the Greeks to some spiny plant. These have singular woolly or bristly bodies, which are formed on the crown of the globose plant. The flowers are small and pink in colour. These plants are difficult to grow.

Cactus intortus.

Carnegiea Britton and Rose

One species, South West America and Mexico.

Named in the honour of Andrew Carnegie (1835-1919), steel magnate and philanthropist. This cactus is the the largest of all the cacti, which grows 21 metres high and 61 cm thick with candelabra-like branches. It is also known as 'Giant Cactus' or 'Suwarro'. It is the State Flower of Arizona. Often, the plants attain a great age of 250 years or more. Spines are awl-shaped, brown or greyish. Stem columnar is with 12 to 14 obtuse ribs. Flowers are greenish on the outer side and white on inner side. The flowers have numerous stamens with yellow anthers.

Carnegiea gigantea.

Cephalocereus Pfeiff.

49 species, Florida to Brazil and Mexico.

The name is derived from the Greek word '*kephale*' meaning a head and 'cereus' a genus of cactus. A woolly head is produced when the plant is of flowering size. It is also known as 'Old Man Cactus'. These are extremely popular with plant collectors. Plants are tall, erect, columnar and generally woody or hairy at the areoles. Flowers are small, nocturnal, fleshy, bell or funnel-shaped.

Cephalocereus hoppenstedtii, Cephalocereus senilis.

Cereus Mill.

50 species, South America and West Indies.

The name is derived from the Latin word '*cereus*' meaning a wax taper, in allusion to the shape of some of the species of this cactus. The plants under this genus may be tall trees, shrubs or climbers. They can grow erect or spread out with ribbed branches. The flowers are very delicate and attractive, funnel-shaped, some being elongated. They bloom mostly in the darkness of the night.

Cereus aggregates, Cereus alacriportus, Cereus azureus, Cereus chalybaeus, Cereus fulgidus, Cereus jamacaru, Cereus peruvianus, Cereus tetragonus.

Chamaecereus Britton and Rose

One species, Argentina.

The name is derived from the Greek word *'chamai'* meaning on the ground and *'cereus'* a genus of cactus, an allusion to its prostrate habit. Branches are cylindrical and creeping.

Chamaecereus silvestrii.

Cleistocactus Lemaire

30 species, South America Molly Andes.

The name is derived from the Greek word *'kleros'* meaning chance and *'dendron'* a tree. Plants have slender and erect stems and are usually very spiny. The spines are usually bright in colour, and flowers are small and bright in colour.

Cleistocactus baumanii, Cleistocactus straussi.

Cochemiea Walton

Five species, lower California.

The name is derived from the name of an Indian tribe, which once lived in lower California, the home of these plants. These are cylindrical, tubercled plants with tubular slender flowers.

Cochemiea maritime, Cochemiea pondii, Cochemiea poselgeri, Cochemiea setispina.

Coryhantha Lem.

64 species, South West America, Mexico and Cuba.

The name is derived from the Greek word *'koryphe'* meaning a summit and *'anthos'* a flower. The flowers are produced on the top of these globular or cylindrical cacti. All species possess large flowers.

Coryhantha aggregate, Coryhantha deserti, Coryhantha palmeri, Coryhantha radians, Coryhantha recurvata.

Crptocereus Alexander

Two species, Mexico and Central America.

These are night blooming and climbing plants with deeply lobed stems. Flowers are attractive, highly scented and last for a short duration.

Crptocereus anthonyanus.

Deamia Britton sad Rose

One species, Mexico and Columbia.

Named after Charles C. Deam, an American botanist. These are climbing plants with numerous needle-like spines. Flowers are funnel-shaped, large and diurnal.

Deamia testuda.

Discocactus Pfeiffer

Nine species, Brazil and Paraguay.

The name is derived from the Greek word 'diskas' meaning a disk and 'cactus', from the shape of these plants, which are globe-shaped or flattened endwise. Plants are small, globular and flowers are nocturnal.

Discocactus hartmannii.

Dolichothele Britton and Rose

Three species, Texas and Central Mexico.

The name is derived from the Greek word '*dolichos*' meaning long and 'thele', a nipple, an allusion to the elongated tubercles or small aboveground tubers of these cacti. These are globular plants with long tubercles but they are not grooved. The flowers are very large with long tube.

Dolichothele longimamma, Dolichothelesphaerica, Dolichothele uberiformis.

Echinocactus Link and Otto.

Ten species, South America and Mexico.

The name is derived from the Greek word '*echino*' meaning a hedgehog; these cacti are like hedgehogs in appearance. These plants are attractive with globular or small columnar ovoid stems that forms plump cacti. These plants have many strong spines. Flowers are pretty, usually yellow in colour.

Echinocactusgrusonii, Echinocactus horizonthalonius, Echinocactus ingens, Echinocactus ingens visnaga.

Echinocereus Engelm

75 species, South America and Mexico.

The name is derived from the Greek word '*echinos* meaning a hedgehog. This genus of cactus has a spiny fruit, which differentiates it from *Cereus*. Plants have low growth with erect or occasionally prostate, cylindrical branches growing singly or forming groups. Flowers are spiny, large, beautiful, bell or funnel-shaped and diurnal, but not always closing at night.

Echinocereus blanckii, Echinocereus chloranthus, Echinocereus dasyacanthus, Echinocereus engelamanii, fitchii, Echinocereus fendeleri, Echinocereus knippelianus, Echinocereus pectinatus, Echinocereus pentalophus, rigidissimus, Echinocereus triglochidiatus, Echinocereus viridiflorus.

Echninomastus Britton and Rose

Ten species, South America and Mexico.

The name is derived fromthe Greek word *'echinos'* meaning a hedgehog and *'mastos'* a breast. The tubercles of this cactus are spiny. These are small, globular or short cylindrical plants with short spines and many ribs. The ribs are notched in the warts and the areoles are woolly.

Echninomastus macdowellii.

Echinofossulocactus Lawrence

32 species, Mexico.

The name is derived from the Greek word *'echinos'* meaning a hedgehog and *'fossula' a* groove. The plants are globular, small, with many ribs and spines. The areoles are very distinct. Flowers are small and short-tubed.

Echinofossulocactus lloydii, Echinofossulocactus multicostatus, Echinofossulocactus phyllacanthus, Echinofossulocactus violaciflorus, Echinofossulocactus wippermannii, Echinofossulocactus zacatecasensis.

Echinopsis Zucc.

36 species, South America.

The name is derived from the Greek word *'echinos'* meaning a sea urchin (also hedgehog) and *'opsis'* meaning like, from the resemblance of these round, spiny cacti to sea urchins. These popular and attractive plants are short and cylindrical or globular in shape. They produce the most beautiful flowers. Established plants produce off-sets and are often rooted on the plant.

Echinopsis campylacantha, Echinopsis decaisneana, Echinopsis eyriesii, Echinopsis hamatacanta, Echinopsis kermesiana, Echinopsis mirabilis, Echinopsis multiplex, Echinopsis rhodotricha, Echinopsis turbinate, Echinopsis tubiflora.

Epiphyllum Haw.

21 species, Mexico to tropical South America and West Indies.

The name is derived from the Greek word '*epi*' meaning upon and '*phyllum*' meaning a leaf. It was originally thought that the flowers of these cacti were on the leaves, which ultimately proved to be flattened stems. They grow on trees in the wild state. They have flattened leaf-like stems. Plants are large, spineless, upright and branched. The flowers are very large, beautiful and trumpet-shaped. There are several hybrids of this cactus with a wide range of colours. Plants are floriferous and each flower lasts for two days.

Epiphyllum ackermanii, Epiphyllum crysocardium, Epiphyllum crenatum, Epiphyllum hybrid, Epiphyllum oxypetalum.

Epiphyllopsis Backeb. and F. N. Knuth

One species, Brazil.

This is a bigeneric hybrid from *Schlumbergera gaetneri* and *Rhipsalidopsis rosea*. These are free-flowering bushy plants with star-shaped rosy red flowers.

Epiphyllopsis graeseri.

Epithelantha Weber ex. Britton. and Rose

Three species, America and Mexico.

The name is derived from the Greek word '*epi*' meaning upon, '*thle*' meaning a nipple and '*anthes*' meaning a flower. A pretty plant with closely set white spines. Plants are globular, depressed at the top and up to 12 cm in diameter. The flowers of this globular cactus are borne on tubercles.

Epithelantha micromeris.

Erythrorhipsalis Berger

One species, South Brazil.

These plants differ from *Rhipsalis* by the minute areoles, which are furnished with bristle-like spines.

Erythrorhipsalis pilocarpa.

Escobaria Britton and Rose

20 species, South West America and Mexico.

This genus is named after two Mexicans, Romulo and Numa Escobar. These plants grow in clusters. The tubercles are grooved and small flowers are borne on these grooved tubercles.

Escobaria roseana, Escobaria shaferi.

Escontria Rose

Three species, Mexico.

This tree like cactus is named after Don Blas Escontria, a Mexican of distinction. Plants freely produce branches from a short trunk. Ribs are few but prominent. Spines are light coloured. Flowers are small and yellow in colour.

Escontria chiotilla.

Espostoa Britton and Rose

11 species, Tropical South America.

This is a popular and attractive cactus. Plants are in columns, hairy or woolly. They develop a cephalium from the central axis. Flowers are small, pinkish and hidden in wool.

Espostoa dantwitzii, Espostoa lanata, Espostoa plumosa.

Ferocactus Britton and Rose

35 species, South West America and Mexico.

The name is derived from the Latin word '*ferox*' meaning ferocious, savage, in allusion to the horrible spines with which this plant is armed. These are strong cacti with heavy spines. Tine stems are round or cylindrical. Areoles are large. The flowers are borne on top of the plant. The growth of these plants is very slow.

Ferocactus acanthodes, Ferocactus cavillei, Ferocactus echidne, Ferocactus latispinus, Ferocactus pringlei, Ferocactus wislizenii.

Frailea Britton and Rose

12 species, Andes and Subtropical America.

These are small plants, which resemble *Mammillaria*, but the flowers are borne on the tubercles and are usually cleistogamous.

Frailea grahliana, Frailea schlinzkyana

Gymnocalycium Pfeiff.

60 species, Subtropical South America.

The name is derived from the Greek word '*gymnos*' meaning naked and '*kalyx*' meaning a bud because the flower buds of these plants are naked. These are spherical in shape and extremely beautiful with a chin-shaped extension below the areoles. The spines are spreading and sometimes recurved. The most important character of this genus is the smooth naked flower buds and flower tubes. Plants of many species are attractively coloured.

Gymnocalycium bodenbenderianum, Gymnocalycium bruchii, Gymnocalycium damsii, Gymnocalycium denudatum, Gymnocalycium gibbosum, Gymnocalycium kurtzianum, Gymnocalycium labodensis, Gymnocalycium leeanum, Gymnocalycium mihanovichii, Gymnocalycium multiflorum, Gymnocalycium platense, Gymnocalycium saglione.

Haageocereeus Backeb.

50 species, Peru and Chile.

Plants of this genus are closely allied to Borzi-cactus, but in this genus the flowers are larger with a longer tube and are borne near the top of the plants.

Haageocereeus acranthus, Haageocereeus chosicensis, Haageocereeus decumbens, Haageocereeus versicolor.

Hamatocactus Britton

13 species, South America and Mexico.

The name is derived from the Latin word '*hamatus*' meaning hooked, plus 'cactus', a mistaken allusion, for the spines of this genus are straight. Plants are round or elongated and ribbed. The flowers are funnel-shaped with a short slender tube.

Hamatocactus setispinus.

Harrisia Britton

13 species, Florida and West Indies.

The name is given after William Harris (1860-1920), Superintendent of the Public Gardens, Jamaica. These are popular plants with slender tall stems and 5-11 ribs. Spines are variable. Flowers are large, showy, funnel-shaped and nocturnal.

Hamatocactus bonplandii, Hamatocactus eriophora, Hamatocactus fragrans, Hamatocactus gracilis, Hamatocactus martini, Hamatocactus tortuosa.

Hatiora Britton and Rose

Four species, South East Brazil.

The name is given in the honour of Thomas Hariot, a 16th century botanist. These are epiphytic, spineless and branched cacti. Stems are green or purplish in colour. Flowers are star-like and variously coloured.

Hatiora bambusoides, Hatiora cylindrical, Hatiora salicorniodes.

Heliocereus Britton and Rose

3-4 species, Mexico and Central America.

The name is derived from the Greek word *'helios'* meaning the Sun and *'cereus'* a cactus, because of the desert habitat of these procumbent Mexican plants. These are beautiful and bushy cacti. Stems are usually procumbent and branches are strongly angled or ribbed. They have scaly or spiny ovaries. Flowers are diurnal and extremely beautiful.

Heliocereus speciosa.

Hylocereus Britton and Rose

25 species, Mexico to Peru.

The name is derived from the Greek word *'hule'* meaning wood and *'cereus'* a cactus. These are climbing cacti with snake-like branches. Flowers are nocturnal and attractive.

Hylocereus undatus.

Lemaireocereus Britton and Rose

25 species, Central America to Venezuela and Colombia, West Indies.

The name is given after Charles Lemaire (1801-1871), a french specialist in cacti. These plants grow extensively and branch from just above the ground. They are mostly tall and erect-stemmed. Flowers are white and open during the day.

Lemaireocereus beneckei, Lemaireocereus marginatus, Lemaireocereus pruinosus.

Lepismium Pfeiff.

One species, Brazil.

The name is derived from the Greek word *'lepis'* meaning a scale, referring to the small scales subtending the areoles. These are epiphytic or lithophytic plants, much branched, flat stemmed. Flowers are with short tubes, white or pink in colour.

Lepismium cruciform.

Leuchtenbergia Hk. f.

One species, Mexico.

A peculiar cactus with long tap roots. The tubercles are finger-shaped, stems woody. The areoles are at the tips of the tubercles. Flowers are long lasting, large, fragrant, yellow.

Leuchtenbergia principis.

***Lobivia* Britton and Rose**

75 species, Argentina, Bolivia and Peru.

The name comes from an anagram of Bolivia where these cacti are found. These are beautiful globular plants with strong pines. Flowers are funnel-shaped, up to 10 cm across, with short tubes, diurnal and brightly coloured.

Lobivia aureus, Lobivia binghamiana, Lobivia bruchii, Lobivia corbula, Lobivia cumingii, Lobivia cylindrical, Lobivia cinnabaina, Lobivia famatimensis, Lobivia Formosa, Lobivia hertrichiana longispina, Lobivia mistiensis, Lobivia pentlandii.

***Lophocereus* Britton and Rose**

Four species, South-west America and Mexico.

These are slow-growing plants with dull green erect stems, branching from the base. These plants have 5 to 7 ribs, and large areoles with 4 to 7 spines. The flowers are nocturnal and pinkish in colour.

Lophocereus schotii.

***Lophophora* Coult.**

1-3 species, South America and Mexico.

The name is derived from the Greek word *'lophis'* meaning a crest and *'phoreo'* meaning to bear, from the tufts of hair borne on the equivalent of leaf axial of these plants. These are rare and graceful plants. These plants are spineless and are round in shape, with flat ribs, bearing tubercles with the areoles, with hair.

Lophophora williamsii.

***Machaerocereus* Britton and Rose**

Two species, Lower California.

The name is derived from the Greek word *'machair'a'* meaning a dagger and *'cereus'* a cactus, a reference to the spines. Plants are prostrate with large, felted and very spiny areoles. Flowers are diurnal and yellow or purple in colour.

Machaerocereus eruca.

***Mammillaria* Haw.**

300 species, South-west America to Columbia, Venezuela and the West Indies.

The name is derived from the Latin word *'mammillo'* meaning a nipple, referring to the form of this cactus. This is a complicated genus, hemispherical in form with the surface broken into prominent tubercles that have conspicuous woolly hair and bristles. The flowers are small bell-shaped that appear in a ring around the top of the plant and usually last for some days. Besides these pretty flowers, they produce attractive fruits.

Mammillaria blossfeldiana, Mammillaria boscasana, Mammillaria bombycina, Mammillaria camptotricha, Mammillaria candida, Mammillaria carnea, Mammillaria celsiana, Mammillaria compressa, Mammillaria decipiens, Mammillaria elongate, Mammillaria erythrosperma, Mammillaria fragilis, Mammillaria gracilis, Mammillaria hahniana, Mammillaria heyderi, Mammillaria kunzeana, Mammillaria kewensis, Mammillaria microcarpa, Mammillaria microhelia, Mammillaria plumosa, Mammillaria parkinsonii, Mammillaria pusilla, Mammillaria prolifera, Mammillaria rhodantha, Mammillaria saetigera, Mammillaria schelhasei, Mammillaria wildii.

***Mediolobivia* Beckeb.**

syn: Rebutia, 41 species *Rebutia arenacea, R. minuscule, R. spegazziniana, R. caniguerali, R. muscular, R. steinbachi, R. fiebrigii, R. narvaecensis. R. steinmannii, R. fabrisii, R. neocumingii, R. marsoneri, R. pseudodeminuta, R. krainziana, R. mentosa, R. pygmaea,* Argentina.

This is an intermediate cactus between *Lobivia* and *Rebutia*. Plants grow in tufts with globular stems. The flowers are large and yellow in colour with a slender, hairy and scaly tube.

Mediolobivia dunrsmaiana, Mediolobivia elegans.

***Melocactus* Link and Otto**

30 species, West Indies, Tropical America.

These are small globules or cylindrical plants, may be solitary or clustered with 9-20 ribs and clusters of spines. Flowers are small and pink in colour.

Melocactus bahiensis, Melocactus broadwayi, Melocactus communis, Melocactus intortus, Melocactus macrocanthus, Melocactus matanzanus, Melocactus melocactoides, Melocactus neryi, Melocactus townsendii.

***Mila* Britton and Rose**

12 species, Peru.

These are small plants with cylindrical stems that grow in clumps. Flowers are small, yellow in colour, and arise from the center of the plant.

Mila caespitose.

Monvilea Britton ad Rose

12 species, Tropical South America.

These have long slender stems, more or less prostrate, with nocturnal flowers.

Monvilea cavendishii, Monvilea rhodoleucantha.

Myritillocactus Console.

Four species, Mexico and Guatemala.

The name is derived from the Latin word '*myrtillus*' meaning small myrtle. The allusion is to the small myrtle-like fruit, not unlike blueberries, of this plant. These are large, stout stemmed, much-branched plants with a few ribs. The flowers are small and diurnal.

Myritillocactus geometrizans.

Neobesseya Britton and Rose

Six species, South-west Amaica and Mexico.

The name is given after Dr. Charles Bessey (1845-1915), a professor of botany, University of Nebraska. These are globular plants with irregular or spiralled grooved tubercles. Flowers are yellow or pink in colour.

Neobesseya missouriensis.

Neocardenasia Backeb.

2 species, Bolivia.

These are stout, columnar plants with deep ribs. Stems are bluish-green in colour. Areoles are set with straw-brown tufts, and the spines are brown and needle-like.

Neocardenasia herzogiana.

Neoporteria Britton and Rose

30 species, Chile.

The name is given in the honour of Charles Porter of Chile, a well known entomologist. Plants are round and hairy at the crown. Flowers are pink, red or yellow in colour, arising from the centre of the plants.

Neoporteria chilensis, Neoporteria fusca, Neoporteria subgibbosa.

Neoraimondia Britton and Rose

Four species, North and West Peru to North Chile.

The name is given in the honour of Antonio Raimondi (1825-1890), a great geographer and naturalist in Peru. These are tree-like cacti with very stout sterns. Areoles have large cushions of thick brown felt, and spines are up to 25 cm long.

Neoraimondia macrostibas.

Nopalea Salm-Dyck

syn: *Opuntia coccinellifera*, mexico, 1 spp

It drives the name from the Mexican name nopal *Dactylopius coccus*, scale insect, a primarily sessile parasite, lives on cacti from the genus *Opuntia*, feeding on moisture and nutrients in the cactus sap.

These plants are almost tree like and bear few glochids. The flowers are red, differing from those of *Opuntis* in that the perianth segments are erect and form a sort of tube. The tile protruding stamens are longer than the petals.

Nopalea coccinellifera.

Nopalxochia Britton and Rose

Three species, Mexico to Peru.

The name is derived from the Aztec name for these epiphytic cacti. Stems are flat, spineless and jointed. Flowers are short tubed, large, pink or red in colour with short tubes.

Nopalxochia phylllanthoides.

Notocactus Backeb. and F. M. Knuth

25 species, Subtropical South America.

In these plants, the stems are slender and trailing in habit. There are numerous ribs, and small areoles. Some plants are strong-growing and erect in habit. These are mostly night-blooming cacti with large or medium flowers.

Notocactus apricus, Notocactus concinnus, Notocactus floricomus, Notocactus haselbergii, Notocactus graessneri, Notocactus leninghausii, Notocactus muricatus, Notocactus ottonis, Notocactus pampeonus.

Nyctocereus Britton and Rose

Seven species, Mexico and Central America.

The name is derived from the Greek word '*nyktos*' meaning night and '*cereus*' a cactus because they are night-blooming cacti. These are erect or climbing plants with cylindrical stems, and are heavily ribbed. Areoles have white wool.

Nyctocereus serpentinus.

Obregonia **Fric.**

One species, Mexico.

It is a very attractive free flowering cactus that resembles an artichoke in appearance. The tubercles are leaf-like, and arranged spirally with areoles at the top. These bear wool and pretty spines when young, which disappear with age.

Obregonia denegrii.

Opuntia **Mill.**

250 species, America.

The name is derived from the Greek name for a different plant, which grew around Opuntium in ancient Greece, 'Prickly Pear'. These include both large and small plants. The stems are jointed where the joints may be flat, cylindrical, clavate or globose, bearing areoles and glochids. Several species bear rudimentary leaves, which disappear with increasing age. They produce colourful flowers in large numbers.

Opuntia ozurea, Opuntia bigelovii, Opuntia clavarioides, Opuntia cylindrica, Opuntia diademata, Opuntia elata, Opuntia floccosa, Opuntia fulgida, Opuntia tunicata, Opuntia kleinae, Opuntia ramosissima, Opuntia spinosior, Opuntia microdasys, Opuntia rufida, Opuntia santarita, Opuntia subulata, Opuntia vestita, Opuntia polyantha, Opuntia versicolor, Opuntia rufinesquei, Opuntia robusta, Opuntia monacantha.

Oreocereus **Riccob.**

Oreocereus Riccob.: genus of about 6 species. Mainly columnar, perennial cacti from mountains region in South America.

Andes.

These are also known as 'Mountain Cactus'. These are graceful plants, covered with longish white silky hair, bearing numerous strong, yellow spines that are 5 cm long. They branch freely from the base. The flowers are small and dark red in colour.

Oreocereus celsianus, Oreocereus fossulatus, Oreocereus hendriksenianus, Oreocereus ritteri, Oreocereus trolli.

Orya **Britton aod Rose**

One species, Peru.

The name is derived from the name of a village in Peru where this species is found. These are little, round-shaped attractive plants. The ribs are divided into low tubercles with numerous spines, and yellowish areoles. Flowers are funnel-shaped, short-tubed and pale pink in colour with yellow at the base.

Orya peruviana.

***Pachycereus* Britton and Rose**

Five species, Mexico.

The name is derived from the Greek word *'pachys'* meaning thick and *'cereus' a'* cactus. These plants have very stout stems. The plants are tall, handsome, and branching at the base. These plants produce white, yellowish or brownish wool from areoles on the top of the plants.

Pachycereus chysomallus, Pachycereus pringlei.

***Parodia* Speg.**

35 species, Tropical and Subtropical South America.

These are small, rare and lovely plants with globular or elongated stems.

Parodia chrysacanthion, Parodia mutabilis, Parodia microsperma, Parodia nivosa, Parodia sanguiniflora, Parodia stuemeri.

***Pelecyphora* Ehrenb.**

Two species, Mexico.

The name is derived from the Greek word *'pelekos'* meaning a hatchet and *'phores'* meaning to bear. The flattened tubercles of these plants are thought to be hatchet-shaped. The plants are small and grow in clusters. The areoles are very long and narrow. Flowers are purple in colour.

Pelecyphora aselliformis.

***Peireskia* Mil.**

20 species, Mexico to Tropical South America and the West Indies.

These leafy cacti are named after Nicholas Claude Fabry de Peirese (1580-1637), a French naturalist. These plants have the appearance of ordinary shrubs. The stems are long and the leaves are somewhat thick, rubber-like. Areoles are in the axils and are large, woolly and spiny. Flowers are solitary or in cluster.

Peireskia aculeate, Peireskia bleo, Peireskia godseffiana, Peireskia grandifolia.

***Pseudoespostoa* Backeb.**

16 species, syn: *Espostoa*

Peru.

These are dark green columnar plants with rose-brown tipped spines, and densely covered with woolly hair.

Pseudoespostoa melanostele.

Pseudorhipsalis Britton and Rose

Pseudorhipsalis alata, P. ramulosa

Mexico.

These are much-branched, prostrate or hanging epiphytic plants. There are numerous flowers, borne on lateral areoles, and the outer petals are yellow and inner petals are white in colour, and lemon scented.

Pseudorhipsalis macrantha.

Pyrrhocactus Backeb. and F. M. Kunth

Six species, Argentina and Chile.

This is a globular, light green coloured cactus with many ribs, notched into low tubercles, with a prominent chin below the areoles. Flowers are funnel-shaped.

Pyrrhocactus tuberisulcatus.

Rathbunia Britton and Rose

Four species, Mexico.

The name is given after Dr. Richard Rathbun (1852-1918), Assistant Secretary of the Smithsonian Institute, Washington DC. These are bushy plants with slender stems and few but prominent ribs. The spines are awl-shaped and whitish. Flowers are red and diurnal.

Rathbunia alamosensis.

Rebutia K. Schum.

Four species, Bolivia and Argentina.

The name is given after P. Rebut, French dealer in cacti, c. 1900. These are popular and attractive cacti. Plants are small and usually freely clustered. Flowers are lovely, few, and funnel-shaped, and formed around the sides of the plant. These are very short-lived plants compared to other cacti. These are very useful for bowl gardens.

Rebutia aureiflora, Rebutia deminuta, Rebutia fiebrigii, Rebutia minuscula, Rebutia pseudominuscula, Rebutia salmonea, Rebutia senilis, Rebutia spinosissima, Rebutia xanthocarpa.

Rhipsalidopsis Britton and Rose

Two species, South Brazil.

These are shrubby, erect or pendant plants with small areoles, sometimes with bristles. Flowers are delicate, small, wide open, regular, and rosy pink in colour.

Rhipsalidopsis rosea.

Rhipsalis **Gaertn.**

60 species, West Indies, Brazil and Argentina.

The name is derived from the Greek word *'rhips'* meaning wicker-work referring to the interlacing twigs on this epiphytic cacti. These plants are branched, with cylindrical, angular or flat leaf like stems with many joints. The areoles are scattered along the margins of the joints and stems. They often carry woolly hair or bristles. These are good for planting in hanging baskets.

Rhipsalis anceps, Rhipsalis casutha, Rhipsalis cereuscula, Rhipsalis elliptic, Rhipsalis houlletiana, Rhipsalis mesembryanthemoides, Rhipsalis pilocarpa, Rhipsalis salicornioides, Rhipsalis paradoxo, Rhipsalis warmingiana.

Schlumbergera **Lemaire**

Five species, Brazil.

The name is given after Frederick Schlumberger, Belgian horticulturist, c. 1900. A popular house plant with flat jointed stems, branching considerably. Large attractive flowers produced in terminal clusters. Fruits are pear-shapcd and coloured. These plants are suitable for hanging baskets.

Schlumbergera gaertneri, Schlumbergera russelliana.

Selenicereus **Britton and Rose**

25 species, South America to Central America, North coast of South America.

The name is derived from the Greek word *'selene'* meaning the moon and *'cereus'* a cactus. Also knwon as the 'Queen of the Night'. The flowers of cacti are mostly nocturnal. These plants are slender, climbing type, ribbed or angled, producing aerial roots. The flowers are probably the finest and largest of all Cactaceae.

Selenicereus boeckmannii, Selenicereus grandiflorus, Selenicereus macdonaldiae, Selenicereus pteranthus, Selenicereus spinulosus.

Stenocactus **A. Berger**

30 species, South America.

These are small plants with many thin ribs. Flowers are small, short-tubed and variedly coloured.

Stenocactus multicostatus.

Thelocactus Britton sad Rose

17 species, South America and Mexico.

The name is derived from the Greek word *'thele'* meaning a nipple and cactus. The ribs have·nipple-like appearance. Plants are globular, some-times very spiny. The flowers are nocturnal, large, funnel-shaped, mostly white in colour and highly perfumed.

Thelocactus bicolour, Thelocactus candicans, Thelocactus gielsdorf-ianus, Thelocactus macrogonus, Thelocactus nidulans, Thelocactus schickendandantzii, Thelocactus spachianum, Thelocactus wagnerianus.

Toumeye Britton and Rose

Toumeya, 15 species, syn: *Sclerocactus papyracanthus*

New Mexico.

These are small globular or cylindrical plants with spiral ribs, and white coloured spines. Flowers have silky white petals.

Toumeye papyracantha.

Trichocereus Riccob.

40 species, Subtropical and Temperate South America.

The name is derived from the Greek word *'thrix;'* meaning a hair and *'cereus'* a cactus. The flowering parts of these cacti are hairy. These are columnar ribbed plains. Flowers are nocturnal, large, white in colour and fragrant.

Trichocereus auricolor, Trichocereus bridgesii, Trichocereus chiloe-nsis, Trichocereus coquimbanus, Trichocereus macrogonus, Trichocereus passacana, Trichocereus schickendantzil, Trichocereus spachianus, Trichocereus strigosus, Trichocereus thelegonus.

Utahia Britton and Rose

Utahiasyn: Pediocactus, 6-11 species.

Utah.

These are small globular cacti with white woolly tops. Ribs are very prominent with 13-15 whitish coloured radial spines and 3-4 dark brown central spines. Flowers are yellow in colour.

Utahia sileri.

Weberocereus Britton and Row

Three species, Central America.

These epiphytic cacti are allied to *Selenicereus*. The stems are slender and trailing. Flowers are nocturnal and white, pink or yellow in colour.

Webercereus biollevi, Weberocereus panamensis.

Wilcoxia Britton and Rose

Six species, South-west America and Mexico.

The name is given after Gen. Timothy E. Wilcox, U.S. Anny, a keen student of plants, c. 1900. These are interesting and pretty free-flowering plants with large tuberous roots and slender weak stems. The diurnal flowers are reddish-purple in colour, and up to 5 cm in diameter.

Wilcoxia poselgeri, Wilcoxia schmollii.

Zygocactus K. Schum

One species, Brazil, Zygocactus, syn: *Schlumbergera trumcatus*

The name is derived from the Greek word '*zygos*' meaning a yoke and *cactus*, from the way in which the stems are jointed. Plants are built up of short, flat, thin joints with small areoles. The spines are minute. The joints are spineless and glossy green in colour. This is a very attractive plant and commonly known as 'Christmas Cactus'. There are many varieties of this cactus with flowers of different colours.

Zygocactus truncactus.

Some very important types of ornamental succulent plants are described below. The species mentioned under each genera are interesting and attractive.

Adenium Roem and Schult.

Apocyaceae, 15 species, Tropical and Subtropical Africa, Arabia.

The name is derived from Aden. These are evergreen succulent shrubs 1-1.5 m long with thick stems and rather fleshy leaves. Clusters of starry white flowers with pink-edged petals appear at the end of the stems.

Adenium obesum, Adenum multiflorum.

Aoenium Webb and Berth.

Crassulaceae, 40 species, Atlantic Island, Mediterranean region, Abyssinia, Arabia.

These are closely allied to *Sempervivum* but the group is marked by technical characters such as calyx with 6-12 teeth or lobes, and carpels partially sunk in the calyx tube. Alternate, entire, fleshy, in rosettes, sessile or short-stalked leaves. Infloresence is paniculated. Flowers are yellow, whitish or pink in colour.

Aoenium arboreum, Aoenium caespitosum, Aoenium ciliatum, Aoenium domesticum, Aoenium tabulaeforme, Aoenium nobile, Aoenium undulatum.

Agave L.

Agavaceae, 300 species, from United States to Tropical South America.

The name is derived from the Greek word *'agaros'* meaning admirable. These are very handsome plants with usually stiff, spiny and often succulent long leaves in rosettes or basal. Some plants have variegated leaves. Flowers are funnel-shaped, greenish-yellow or red in colour, and are borne on 10 to 12 metres high spikes when plants attain 10-60 years of age.

Agave americana, V angustifolia var. *marginata, Agave decipiens, Agave filamentosa, Agave leopoldi, Agave sartori.*

Aloe L.

Liliaceae, 275 species from Tropical and South Africa, 42 species from Madagascar, 12-15 species from Arabia.

The name *Aloe* is derived from the Arabic word *'olloch"*. These plants are small or tree-like. Leaves are stiff, spiny and toothed, usually crowded in rosettes or along the end of the stem. Flowers are showy, red or yellow in colour, and often paler striped. Some species excite astonishment by producing flowers 6-10 m above the ground.

Aloe africana, Aloe arborescens, Aloe aristata, Aloe brevifolia, Aloe variegate, Aloe candelabrum, Aloe claviflora, Aloe ferox variegate, Aloe latifolia, Aloe striata, Aloe succotrina, Aloe variegata

Apicra Willd.

Liliaceae, 10 species, South Africa.

These are *Aloe*-like succulents with short 10-30 cm long stems, crowded or spirally arranged leaves. The flowers are greenish in colour and tubular in racemes.

Apicra aspea, Apicra deltoidea, Apicra dodsoniana, Apicrae egregia, Apicra foliolosa, Apicra pentagona, Apicra Apicra spiralis.

Bryophyllum Salisb.

Crassulaceae, 20 species, Warm regions.

These succulents have vegetative buds on the edges of the leaves, which need only to be laid on damp sand to start new plants. The stem of these succulents is upright, simple and sometimes branching. The root system is fibrous, and the leaves are opposite, petioled and simple or pinnately compound. Inflorescence is cymose or paniculated with nodding flowers. Flowers are red to violet or grey-green in colour.

Bryophyllum diagremoutianum, Bryophyllum tubiflora.

Caralluma R. Br.

Asclepiadaceae, 110 species, Africa, Mediterranean region to Myanmar.

These are leafless succulents with angled stems, allied to *Stapelia*. The flowers are usually in clusters.

Caralluma burchardii, Caralluma eurapaea, Caralluma lutea, Caralluma mammillaris, Caralluma nebrownii, Caralluma suaveolens.

Ceropegia L.

Asclepiadaceae, 160 species, Canary Islands and South Africa, Madagascar, Tropical and Subtropical Asia (Arabia to New Guinea).

This generic name is in allusion to the waxen appearance of the flower clusters on these trailing plants. The leaves are opposite, sometimes lacking. The flowers are tubular, mostly in axillary clusters, and white, green or purple in colour.

Ceropegia ampullata, Ceropegia elegans, Ceropegia fusea, Ceropegia sandersonii, Ceropegia woodii.

Conophytum Schwantes.

Aizoaceae, 20 species, South Africa.

Also known as the 'Cone Plant', in allusion to the inverted cone shape of these plants. These are very small succulent plants with abbreviated roots. The plants look either roundish pebble-like or like the other types with two short lobes on top of a pair of leaves. These are stemless when young. Flowers are solitary and white, red, yellow, rosy or magenta in colour.

Conophytum altum, Conophytum batesii, Conophytum calculus, Conophytum diver-gens, Conophytum elishae, Conophytum frutescens.

Crassula L.

Over 300 species, Cosmopolitan, specially South Africa.

The name is in allusion to the somewhat thick leaves of these evergreen dwarf plants, 1-1.2 metres tall. Flowers are in cymes or clusters and white, rose, pink or yellow in colour.

Crassula alstonii, Crassula arborescens, Crassula columnaris, Crassula coruta, Crassula falcate, Crassula lacteal, Crassula lycopodiodes, Crassula picturata, Crassula pyraminidalis, Crassula rupestris.

Cotyledon L.

Crassulaceae, 40 species, South Africa, Arabia.

Also known as 'Navel Wort'.The name *Cotyledon* is in reference to the cup-like leaves of some species of these succulents. The plants have basal or straight leaves, which may be flattened or rounded. Flowers are campanulated in terminal cymes and red or greenish-yellow in colour.

Cotyledon barbeyi, Cotyledon decussata, Cotyledon galpinii, Cotyledon orbiculata, Cotyledon paniculata, Cotyledon reticulata, Cotyledon tuberculosa, Cotyledon undulate.

Dudleya Britton and Rose

Crassulaceae, 60 species, Southwest United States. The name is given in the honour of William Russel Dudley (1849-1911), first professor of botany at Stanford University, California. This is a subdivision of the genus *Echeveria*. Leaves rosettes, often beautifully coloured and sometimes covered with hair. Flowers bell-shaped, in various shades of white and yellow.

Dudleya brauntonii, Dudleya caespitosa, Dudleya candida, Dudleya cotyledon, Dudleya cultrata, Dudleya farinosa, Dudleya grandiflora, Dudleya greenei, Dudleya pulverulenta, Dudleya purpusii.

Duvalia Bonpl (= *Hypocalyptus* Thunb.)

Genus of about 19 species of prostrate of semi erect, mainly leafless, clump forming perennial succulent from hilly lowlands of the Peninsula, E. Africa and South Africa

Leguminosae.

These dwarf succulent herbs are named after Henri Auguste Duval (1777-1818), French medical man. These are leafless plants with thick, angled, and teeth tipped stems. The flowers are in small clusters.

Duvalia cordeoyi, Duvalia polita, Duvalia pubescens, Duvalia radiate, Duvalia reclinata.

Dyckia Schult. f.

Bromeliaceae, 80 species, warm South America.

The name is given after Prince Joseph Salm-Reifferscheid Dyck (1773-1861), author of outstanding books on succulents. These are stemless succulents with stiff, spiny, margined leaves in basal rosettes up to one metre length. Flowers are in racemes or panicles, and yellow or orange in colour.

Dyckia altissima, Dyckia brevifolia, Dyckia fosteriana, Dyckia frigid, Dyckia leptostachya, Dyckia niederleinii, Dyckia rariflora, Dyckia Dyckia remifloria, Dyckia sulphurea.

Echeveria DC

Crassulaceae, 200 species, Southern United States to North Andes.

The name is given after Athanasia Echeverriay Goday, a botanical artist of repute. These are low-growing succulent plants. Fleshy leaves usually broad and flat, commonly making dense rosettes, sometimes highly coloured. Flowers appear in loose spikes or racemes.

Echeveria agavoides. Echeveria bracte-olata, Echeveria derenbergii, Echeveria gibbiflora, Echeveria glauca, Echeveria leucotricha, Echeveria pilosa, Echeveria setosa.

Echidnopsis

Asclepiadaceae, Four species, Africa and Arabia.

These leafless plants allied to *Stapelia*, have stout branching and tubercled stems. The flowers are small, fleshy and yellow or purple-brownin colour.

Echidnopsis cereiformis, Echidnopsis cylindrical, Echidno-psis dammanniana.

Edithcolea N. E. Br.

Asclepisdaceae, Two species, Somalia, Kenya and Tangankya.

These plants are allied to Caralluma. Stem are greyish green, 30 cm long with thorn like teeth. Flowers are generally larger in size, about 10 cm in diameter, and pale yellow in colour.

Edithcolea grandis.

Euphorbia L.

Euphorbiaceae, 200 species, Cosmopolitan, chiefly Subtropical and warm temperate.

Classically supposed to have been named after one Euphorous, physician to the king of Mauritania. These plants show immense diversity of forms. The genus is distinguished by a single pedicellate, pistillate flower without floral envelopes or with only a rudimentary calyx, surrounded by numerous staminate flowers. The inflorescence is surrounded by a more or less cup-shaped involucre called cyathium. The involucre is regular or nearly regular.

Euphorbia bubalina, Euphorbia coerulescens, Euphorbia cooperi, Euphorbia .fimbriata, Euphorbia ferox, Euphorbia fulgens, Euphorbia grandicorms, Euphorbia heterophylla, Euphorbia globosa, Euphorbia meloformis, Euphorbia obesa, Euphorbia ornithoplus, Euphorbia pseudo-globosa, Euphorbia submamillaris, Euphorbia splendens, Euphorbia susannae.

***Faucaria* Schwantes**

Aizoaceae, 36 species, South Africa.

The generic name is derived from the Latin word '*fauces*', the throat, in reference to the paired leaves with tooth like edges of these succulents, which resemble the wide open mouth of a carnivore. These plants are almost stemless, with crowded fleshy leaves and thorn like projections on the sides. Flowers are mostly yellow in colour.

Faucaria albidens, Faucaria felina, Faucaria lupine, Faucaria tigrina, Faucaria tuberculosa.

***Fenestraria* N. E. Brown**

Alizoaceae, Two species, South Africa.

The name is given in allusion to the transparent window-like patch of skin on top of the stem. These are characterised by a tufted stemless habit with upwardly, fleshy, thickened leaves. Flowers are large, showy, and yellow and white in colour.

Fenestraria aurantiaca, Fenestraria rhopalphylla.

***Furcraea* Vent.**

Agavaceae, 20 species, Tropical America.

The name is given after Antoine Fraucois Fourcroy (1700-1809), French chemist and naturalist. Leaves of these plants are 1-2 m long, in the basal rosettes and whitish or greenish in colour with with some spiny foliage. Flowers are greenish-white in terminal panicles on tall scapes often 10-12 m high.

Furcraea bedinghausii, Furcraea gigantea, Furcraea striata, Furcraea longaeva, Furcraea selloa.

***Gasteria* Duval**

Liliaceae, 70 species, South Africa.

The asymmetrically swollen base of the flowers distinguishes this genus from allied Liliaceae. These plants are almost stemless, and the leaves are leathery mostly warted, and mottled in colour. Flowers are small, bell-shaped, greenish-pink in colour, and appear on tall stalks.

Gasteria armstrongii, Gasteria batesiana, Gasteria disticha, Gasteria hybrida, Gasteria liliputana, Gasteria lingua, Gasteria maculata, Gasteria nigricans, Gasteria trigona, Gasteria verrucosa.

***Glottiphyllum* Haw.**

Alizoaceae, 50 species, South Africa.

The leaves of these succulent plants are very thick, soft fleshy and tongue-shaped. They are brittle and easily snapped. Leaves are fleshy and produce lame yellow or white flowers on short stems.

Glottiphyllum album, Glottiphyllum depressum, Glottiphyllum fragrans, Glottiphyllum mariotti.

***Haworthia* Duval**

Liliaceae, 150 species, South Africa.

The name is given after Adrian Hardy Howarth (1768-1833), English authority on succulent plants. These plants are stemless with tongue-shaped to nearly cylindrical leaves. The flowers are inconspicuous, bell-shaped, in long raceme.

Haworthia attenuatta, Haworthia bolusii, Haworthia cuspidata, Haworthia fasciata, Howorthia limifolia, Haworthia maughanii, Haworthia otzenii, Haworthia reinwardii, Haworthia tessellata, Haworthia truncate, Haworthia viscosa.

***Hechtia* Klotzsch**

Bromeliaceae, 35 species, United States to Central America.

The name is given after J. H. G. Hecht, a counsellor to the King of Prussia. Leaves of these plants are long, stiff and spiny toothed in dense rosettes. Flowers are small in terminal panicles.

Hechtia argentea, Hechtia ghiesbreghtii, Hechtia glomerata, Hechtia marnierlapostellei, Hechtia podantha, Hechtia schottii, Hechtia tillandsioides.

***Heurnia* Spreng (= *Huernia* R.Br.)**

Asclepiadaceae, 30 species, South Africa and South Arabia.

These dwarf perennials are named in the honour of Justin Heurnius (1537-1652), a Dutch missionary. Related to *Stapelia*, these have fleshy stems, armed with small teeth on the edges. Flowers are cup-shaped, attractive, yellow to brown in colour, spotted with reddish brown.

Heurnia bicampanulata, Heurnia confusa, Heurnia hystrix, Heurnia hislopii, Heurnia keniensis.

Hoodia Sweet ex. Deene.

Asclepiadaceae, Ten species, South-west Tropical and South Africa.

The name is given after Mr. Hood, an early 19th century succulent cultivator. These are leafless herbs with thick cylindrical angled stems having tubercles armed with spines or bristles. Flowers are yellow, orange or pink in colour and are in small clusters near the tops of stems.

Hoodia bainee, Hoodia gordonii, Hoodia langii, Hoodia pillansii.

Hoya R.Br.

Asclepiadaceae, 200 species, South China, Southeast Asia, Indo-Malay, Australia.

These climbing and flowering evergreen plants are named after Thomas Hoy (1788-1809), 18th century head gardener to the Duke of Northumberland at Syon House, Chiswick, England. These plants produce clusters of wax-like flowers of crimson or white colour. These attain a height of up to 4 m.

Hoya bella, Hoya carnosa, Hoya variegate.

Huerniopsis N.E. Br.

Asclepiadaceae, Four species, South and South-west Africa.

Allied to *Stapelia,* these plants are distinguished from the closely-related *Huernia* by the corolla having intermediate points or lobes at the sinuses between the primary lobes.

Huerniopsis decipiens, Huerniopsis gibbosa.

Kalanchoe Adans

Crassulaceae, 200 species, Tropical South Africa, China and Java, one species, Tropical South America.

The name is derived from the Chinese name for one species of these succulent herbs or sub-shrubs. These are perennial succulents with fragrant flowers. Plants are robust and erect. Leaves are opposite, fleshy, sessile or stalked. Flowers are in many shades including yellow, purple or scarlet.

Kalanchoe beharensis, Kalanchoe blossfeldiana, Kalanchoe campanulata, Kalanchoe eriophylla, Kalanchoe flammea, Kalanchoe longiflora, Kalanchoe marmorata, Kalanchoe pumila, Kalanchoe rhombopilosa.

Lithops N.E. Brown

Aizoaceae, 350 species, South Africa.

This generic name is in reference to the resemblance of these plants to stones. These plants have leaves that are thick, in pairs, and mostly buried as they grow in nature. Flowers are solitary and yellow or white in colour.

Lithops aucampiae, Lithops bella, Lithops bromfieldii, Lithops cristinae, Lithops dorotheae, Lithops fuller, Lithops leslei.

Mesembryanthum L.

Aizoaceae, 350 species, South Africa.

Also known as "Middle-Embryo Flowers', with reference to the position of the embryo. These succulents have creeping habit of growth, and have fleshy leaves producing flowers of great beauty.

Mesembryanthemum albatum, Mesembryanthemum crystallinum, Mesembryanthemum coccineum, Mesembryanthemum cordifolium, Mesembryanthemum variegatum, Mesembryanthemum fuller, Mesembryanthemum gramineum, Mesembryanthemum intransparens, Mesembryanthemum macrophyllum, Mesembryanthemum nudiflorum, Mesembryanthemum setosum, Mesembryanthemum striatum.

Pachyveria Hort. ex Haage and Schmidt

Crassulaceae (*Echeveria* x *Pachyveria* bigeneric hybrid).

These plants show characters of both parent genera.

Pachyveria clavata var. *cristata*, *Pachyveria mirabilis*, *Pachyveria scheidecker*i var. *cristata*.

Pedilanthus Neck ex Poit.

Euphorbiaceae, 20 species, Florida and Mexico to Tropical South America, West Indies.

Also known as 'Slipper Flower', from the appearance of the flowers of this point. These plants have milky juice, its leaves are simple, alternate, and the flowers are in showy cyathia.

Pedilanthus aphyllus, Pedilanthus macrocarpus, Pedilanthus retusus, Pedilanthus tithymalodes.

Piaranthus R. Br.

Asclepiadaceae, 16 species. South Africa.

These are leafless, very dwarf sussulents. Flowers are in pairs or clusters having a single crown.

Piaranthus cornutus, Piaranthus foetidus.

Pleiospilos N. E. Brown

Aizoaceae, 38 species, South Africa.

The leaves of these succulent plants are conspicuously spotted. These stemless plants resemble a piece of rock or granite, and bear two pairs of fleshy leaves. Flowers are yellow in colour.

Pleiospilos bolusii, Pleiospilos dekhanahi, Pleiospilos hilmary, Pleiospilos neilii, Pleiospilos simulans.

Sedum L.

Crassulaceae, 600 species, North Temperate Peru, Mexico.

Also known as 'Stone-crop'. Stems of these plants are woody or fleshy, and prostate. Leaves are opposite or in whorls or cylindrical edge, entire or notched. Flowers are in terminal false umbel or bunch, and white, red, yellow or blue in colour.

Sedum acre, Sedum anacampseros, Sedum anglicum, Sedum brevifolium, Sedum maximum, Sedum oreganum, Sedum rupestre, Sedum spectabile, Sedum spurium.

Sempervium L.

Crassulaceae 25 species, Mountains of South Europe.

These are low growing plants forming clumps or cushions. Leaves are spirally arranged in close rosettes, and are fleshy green, reddish or bluish in colour. These plants have many flowered inflorescence, which are erect, terminal, white, greenish, pink, purple or yellow in colour.

Sempervium arachnoideum var. *glabrescens, Sempervium arenarium, Sempervium borisii, Sempervium calcareum, Sempervium nontanum, Sempervium soboliferum, Sempervium tectorum, Sempervium tectorum,* var. *calcarum,* var. *rubescens,* var. *violescens, Sempervium wulfenii.*

Senecio L.

Compositae, 2000-3000 species, Cosmopolitan.

The name is derived in allusion to the hoary pappus of these herbs. Allied to Kleinia, these plants are often with tuberous rootstock. Leaves are roundish or flat, somewhat cylindrical. Flowers are medium and yellow in colour.

Senecio articulatus, Senecio barbertonensis, Senecio cineraria, Senecio confusa, Senecio fulgens, Senecio gunninsii, Senecio haworthii, Senecio klinghartianus, Senecio longifolius, Senecio paucifolius, Senecio stapeliiformis, Senecio tropaeolofolius, Senecio vestita.

***Stapelia* L.**

Asclepiadaceae, 75 species, Tropical South Africa.

Also known as 'Carrion Flower'. These cactus-like plants are named by Linnaeus after Johannes Bodeus van Stapel, a Dutch physician. These perennial plants have fleshy stems, shnativaped like stag's horn. Flowers large, livid, foetid and of many colours.

Stapelia asteris, Stapelia bufonia, Stapelia grandiflora, Stapelia gigantean, Stapelia hirsuta, Stapelia lepida, Stapelia mutabilis, Stapelia nobiis.

***Tianopsis* Schwantes**

Aizoaceae, Eight species, South Africa.

The name is derived from the resemblance of the flower of these succulents to the Sun. Leaves are very crowded, rosettes and decussste. Flowers are solitary, almost sessile or stalked, medium-sized, and yellow or orange in colour.

Tianopsis calcarea, Tianopsis fulleri, Tianopsis hugoschlechteri, Tianopsis luckhoffii, Tianopsis luderitzii, Tianopsis primosii, Tianopsis setifera, Tianopsis schwantesii.

***Yucca* L.**

Agavceae, 40 species, South U.S.A., Mexico and West Indies.

Also knwon as 'Adam's Needle', these plants are handsome and hardy with rosettes of sword leaves. Flowers are usually borne in erect panicles, up to two metres or more.

Yucca aloifolia var. *tricolor, Yucca australis, Yucca filamentosa, Yucca flaccida* var. *variegata, Yucca gloriosa, Yucca rupicola, Yucca schottii, Yucca whipplei.*

LOCATION OF CACTUS AND SUCCULENTS IN THE GARDEN

A cactus garden can enhance the look of a house. When choosing cacti for the garden, be sure to choose plants that will thrive in the various outdoor conditions. Plant the cacti into the ground in such a way that they can be easily transported indoors during winter. The best place to plant a cactus garden is on a hill, as it allows the garden to drain easily.

When you decide where you want your cactus garden, decide what size and shape you want it to be. Use your imagination and be creative; you can have it in any shape you desire.

Wall-Water Fountains

Cacti and succulents can be used as striking foliage to accompany a wall-water fountain or awning.

Miniature Indoor Desert Scenes

For small apartments or homes with limited space, it is easy to create a miniature desert scene indoors with decorative dish gardens of cacti and succulents. These miniature landscapes offer great interest as they take only a few hours to make and once established, grow almost unattended.

Most Cacti do fairly well as house plants; however they are quite slow growing. Save the sunniest spots in the garden for Cactus plants as they need lots of light. Remember to turn them every few weeks so that they are evenly illuminated. If you are to grow some of your Cacti to flower, or for seeds, then do not move them while in bloom. They are very sensitive to disturbances at this stage and can drop the buds.

Cacti can be more sensitive to Sun burns if exposed to outside conditions during the hot periods of mid-summer. If you are going to put your Cacti outside, they must first be acclimatized. Keep them in shade for a few weeks, and then move them into partial Sun. Do not be in a hurry to scorch them under a hot Sun in July. Give them a month or more to get gradually used to it. If after you put it outside, your cactus starts to acquire a lighter green or tan tint, it is probably sunburned; and they should be moved to some shade.

At the same time, insufficient light will cause the cacti to display pale colouration and result in an elongation and tapering of the growing tips and poor flowering. Sub-tropical epiphytic cacti such as *Schlumbergera*, *Zygocactus* etc. prefer semi-shade during summer for proper growth. All cactus seedlings should be shielded from full Sun until they have developed spines to provide them protection. In tropical areas, plants must be shielded from direct sunlight during the summer and autumn months, using mesh that provides 20-30% shade. During the winter and spring this mesh should be removed.

Succulents do better in partial shade. They are vulnerable to scorching when exposed to direct sunlight. Members of some succulent genera such as Adenium, Euphorbia, Pachypodium, Yucca, Agave, Furcraea and Beaucarnea are well adapted to the ecosystem of the humid tropics and can be grown outdoors in direct sunlight. Haworthia and Gasteria dislike intense sunlight. When mature, most cacti and succulents need full Sun, so they are best cultivated outdoors in warmer climate with a minimum temp of 16°C, in a sunny condition.

Tropical cacti and succulents need a temperature range of 21-32°C. Cacti need a maximum temperature of 7°C at night in the winter and need to be kept completely dry from the end of September to mid-March. Other succulents, generally, prefer higher winter temperature of around 10°C and consequently more regular watering. In Delhi, Kalimpong and Darjeeling where winter temperature goes down to 2°C, cacti enjoy a winter rest and a suspended growth. In eastern part of the country where winter temperature varies between 15-30°C, best growth has been recorded in most species of cacti during this period. The period from March to April, when the average temperature varies between 25-35°C, is the best time for most species of cacti to flower and bear fruits.

RECOMMENDED VARIETIES AND THEIR SALIENT FEATURES

1. *X Ferocalycium* "Sarala"
2. *Gymnocalycium mihanovichii* cv. "Ashoka"
3. *Gymnocalycium mihanovichii* cv. "Dr. J.B. Pattnaik"
4. *Gymnocalycium mihanovichii* cv. "Lingaraj"
5. *Gymnocalycium mihanovichii* var. *friedrichii* cv. "Prof. P.K. Sen"
6. *Gymnocalycium mihanovichii* cv. "Dr. B.B. Sundareshan"
7. *Gymnocalycium mihanovichii* var. *friedrichii* cv. "Bhargavi"
8. *Gymnocalycium mihanovichii* var. *friedrichii* cv. "Arun"
9. *Gymnocalycium mihanovichii* cv. "Ananta" fa.cristata
10. *Gymnocalycium mihanovichii* cv. "Dr. M.S. Swaminathan" fa.cristata
11. *Gymnocalycium mihanovichii* var. *friedrichii* cv. "Ekamra"
12. *Gymnocalycium mihanovichii* cv. "Mayuri" fa. cristata
13. *Gymnocalycium mihanovichii* var. *friedrichii* cv. "Dhauli"
14. *Gymnocalycium mihanovichii* cv. "Krishna"
15. *Gymnocalycium mihanovichii* cv. "Sambhu"
16. *Gymnocalycium mihanovichii* cv. "Mr. Thongler"
17. *Gymnocalycium mihanovichii* cv. "Mangal"

PROPAGATION

Raising from Seed

Dry seeds are available commercially but the best results are obtained by using fresh seeds. It is best to sow the seeds towards the end of winter so that the seedlings are as large as possible by the following winter. Choose a 5 cm pot, a 5 cm deep sand tray or a 13 cm pan for seeds according to the quantity and size of the seed. When watering the compost add some fungicide to stop the seedling from damping off.

Fill a pot almost to the brim with the cacti compost and firm. Scatter the seeds evenly over the surface. Stand the pot in tepid water until the compost surface is moist. Cover the pot thinly with washed grit. Place the pot in a plastic bag or propagator. Leave it in shade at minimum temperature of 19°C or 21-27°C to speed germination.

Pricking Out

Once the seeds germinate (within 2-3 weeks), move the seedlings out of their plastic bags or propagator. Grow them at 10°C in a bright airy position, but not in the Sun, which can scorch them. Wait several months to a year to prick them out. Plant seedlings less than 2.5 cm across in a tray and larger ones in pots.

(I) Prick out seedlings only when they begin to crowd each other or the compost begins to cake. Use a widger to ease out a clump of seedlings with their compost. Take care not to damage the young roots.

(II) Gently separate each seedling from the clump. Fill the pots with almost 5-6 cm of cactus potting compost. Plant one seedling in each pot and firm them gently. Top-dress with a layer of small washed grit. After 3-4 days, water them with a fungicide solution.

Cutting

Cuttings provide a quick and easy means of propagating cacti and succulents. The best time to take cuttings is when a plant starts growing actively, which for most species is in spring. Allow the cuttings to heal and form a callus before they are inserted into the pots. To reduce the risk of rot or disease, dust any wound with a fungicide.

Leaf Cuttings

Succulents such as gasteria and some Haworthia root rapidly from leaf cuttings (in 3-12 weeks). Pull a young healthy leaf away from the parent plant carefully And leave it in a bright, warm, dry place for a day or two until a callus is formed over the wound. Fill a small pot with the cacti compost up to 2.5 cm of the rim. Fill the pot with small washed grit to keep the cutting upright. Set the pot in partial shade at a minimum temp of 18°C. Keep the compost slightly damp. When the new plants appear, separate them and pot them.

1

2

3

4

5

6

7

8

9

10

11

12

Fig. 18. Recommended varieties

Stem Cutting

Choose a healthy, sturdy stem whose removal will not disfigure the mother plant. Cut the stem just above a bud or shoot with a sharp knife. Trim the stem so that the cutting is 5-10 cm long. Remove any lower leaves, if necessary, so that the stem is clean at the base. Fill a small pot up to one-third with cacti potting compost. Cover with small washed grit and firm. The cutting should root in 2-6 weeks.

Division

Cacti and succulents can be propagated readily from stem and tuberous offset by division. Usually the best time to divide these types of cacti and succulents is at the beginning of the growing season in spring or early summer. Some species of succulents, with swollen rootstocks, can be increased in winter while the plants are dormant. To do so, cut off segments from the rootstock and water all newly divided plants sparingly until they become well established and show signs of active growth, otherwise their stems may rot in the damp compost.

Clumping Offsets

Succulents such as haworthia and gasteria and cacti, such as Mammillaria and Echinopsis, produce offsets in a dense clump. Mature offsets which have begun to form their own roots can be detached from the parent to grow into new plant. Once they are potted, they should be left at a minimum temperature of 18°C in Sun or bright shade. They should be watered sparingly until new growth appears.

(I) With a widger, carefully dig out an offset from around the base of the plant. Pull it away gently or cut across the joint with a clean sharp knife. Dust the wound with fungicide and leave to callus for a few days.

(II) Almost fill a suitably sized pot with cactus potting compost. Insert the offset into the compost at the same depth as it was before and firm it gently. Top-dress with small washed grit.

Offset Tubers

Some succulents such as *Senecio* have tuberous roots and form offset tubers just below the soil level and around the parent plant. These can be gently detached or severed with a sharp knife and made to grow independently.

(I) Clear away the top-dressing, if any, and dig carefully with a widger in the potting compost to locate an offset tuber and lift it out of the pot.

(II) Leave the offset in a warm dry place to callus. If the offset has roots, plant it in a pot of cactus potting compost. If it has no roots, use a pot filled with one-third compost and two-third small washed grit.

(III) Cover the compost thinly with small washed grit. Leave it for a few days. Water sparingly until the plant is established.

Division of Rootstocks

This is used for a few clumping plants such as *Sansevieria trifaciata*. When the plant is dormant, cut the rootstock or tuber into pieces, each with its own growing point. Dust the cuts with a proprietary fungicide and allow to callus. Pot up each section and let it grow on in conditions that suit the plant's needs.

Grafting

This process is mainly used with cacti. When grafted, many slow–grafting and difficult species are easier to cultivate and bring into flower. In this way, their growth rate can be increased by ten times.

Split Grafting

(I) Remove 2.5 cm from the top of a healthy stock with a sharp knife. Make a vertical cut 1-2 cm deep into the stem, ensuring that the cut passes the central core tissue.

(II) To create a scion, select a length of healthy stem from the parent plant. Make a straight clean cut through the leaf joint.

(III) Cut away 1 cm of the bottom of the stem section. Be careful to remove the skin from both sides to form a tapered scion. If necessary, trim the scion to the same diameter as the stock.

(IV) Gently insert the prepared scion into the slit at the top of the stock. Check that the central core tissue of the scion is in close contact with that of the stock.

(V) Secure the two parts in position by binding the graft firmly with raffia. Leave it in a bright airy place, at a temperature of least 10°C, for about 2 weeks until new growth is visible.

Flat Grafting

(I) With a clean sharp knife, cut through the stock horizontally about 2.5 cm below the growing point. Remove the top. Check that some central core tissues is clearly exposed.

(II) Bevel the edges of the cut, by removing about 6 mm of the skin holding the scion off the stock when the fleshy stock tissue shrinks after grafting.

(III) Cut through the scion at the base. Bevel 6 mm of the skin from around the cut edge of the scion. Place the cut surfaces of the scion and stock together so that their central core tissues touch.

(IV) Firmly rotate the scions to remove any air pockets. Secure the stock and scion in position with a pair of rubber bands. If the graft is successful, the scion and stock should unite in about 2 weeks.

Side Grafting

This is ideal for grafting very slender stemmed plants, such as *Echinopsis chamacereus*, which often have a very small central core. Cut the stock and scion obliquely, to expose more of the central core tissue. Rest of the procedure is same as for flat grafting. Support it with a stake, if necessary.

At the time of grafting, the temperature should be between 20-25°C, the air should be dry, and the RH should be 60%.

Table 4. Suitable rootstock for different scions

Name of rootstock	Name of scion
a. *Cephalocereus albispinus*	1. *Cephalocereus senilis*
	2. *C. senilis cristatus*
	3. *Cleistocactus strausii* Cristate'
	4. *Espostoa plumosa*
	5. *E.plumosa 'Cristate'*

	6. *Haageocereus bicolor*
	7. *Mammillaria angularis* 'Cristate'
	8. *Notocactus leninghausii*
	9. *Opuntia tunicata*
b. *Cephalocereus* sp.	1. *Cleistocactus strausii'cristate*
	2. *Haageocereus bicolor*
c. *Cleistocactus baumannii*	1. *Aprocactus flagelliformis*
	2. *Chamaecereus silvestrii*
	3. *Mammillaria elongata*
	4. *M. gracilis*
d. *Myrtillocactus geometrizans*	1. *Cephalocereus senilis cristatus*
	2. *Gymnocalycium mihan ovichii rubra*
e. *Opuntia tomentosa*	1. *Opuntia microdasys*

CULTIVATION

Containers

Traditionally clay pots are used as containers. Plastic pots have the advantage of reduction in moisture loss and a more even distribution of root system throughout the available soil.

Compost

For pot plants and small beds, use gritty compost as garden soil is likely to harbour pests and diseases. Most cacti and succulents like a slightly acid soil (pH 5.5-6.5). The bed should be raised at least 15 cm from the ground. Drain off excess water, if any. Compost consisting the following is quite suitable for growing most species:

- One part loamy or sandy loam soil
- One part well-rotten cow dung manure
- Two parts of leaf mould of medium coarseness
- One part of sand
- Small quantities of broken pieces of bone meal

10% BHC should be dusted to keep away the ants and root mealy bugs to some extent. Compost used for seed germination consists of one part of river sand and two parts of leaf-mould. The nutrients available in the compost are adequate for good and healthy growth of most cacti seedlings for about one year. A compost composed of leaf mould and sand in equal proportions is suitable for induction of rooting of cuttings.

Table 5. Recommended propagation techniques for specific genera

Plant Genus	Seed	Stem Cutting	Leaf Cutting	Grafting	Tuber, Bulbil, and Offset Division
Agave	X				X
Aloe	X	X			x
Bryophyllum	X	X	X		X
Cereus	X	x		X	
Ceropegia	X				X
Chamaecereus	X			X	
Echinocactus	X			X	
Echinopsis	X	X		X	X
Epiphytic cacti	X	X		X	
Eriocactus	X			X	
Espotoa	X			X	
Euphorbia	X	X			
Gasteria	X		X		
Graptopetalum	X	X	X		X
Haworthia	X		X		X
Kalanchoe	X	X			X
Lobivia	X			X	
Mammillaria	X	X		X	X
Melanocactus	X			X	
Neoporteria	X				
Notocactus	X	X		X	
Opuntia	X	X			
Rebutia	X			X	
Sansevieria	X	X			
Sedum	X	X	X		
Stapeliads	X	X			
Sulcorebutia	X			X	
Trichocereus	X	X			

Compost for Succulents

It should be porous. A mixture of garden loam, coarse sand, leaf mould, crushed brick and charcoal in the ratio of 2:2:2:1:1 makes ideal compost for most succulents.

Planting into Containers

Plants less than 15 cm across are best grown in pots that are 1" wider than the plants diameter. Larger plants require a pot 5-10 cm wider than their diameter.

Planting in Beds

In frost free areas, cacti and succulents can be planted in any sunny location, as long as the soil is free-draining and not too alkaline or acidic and there is shelter from strong winds. Where rainfall exceeds 635 mm, build a raised bed or a raised sloping mound with a suitable soil or compost. Applying a top-dressing improves drainage and hinders erosion. Harden off indoor plants for a few days in the shade before planting outdoors. For summer plating prepare a bed or a gravel-filled trough.

KEEPING PLANTS HEALTHY

Pruning

Most succulents and some cacti benefit from some shaping and restriction of their growth to make them busier and more compact. Remove any unproductive growth that may impair the vigour of the plant. Prune shrubby succulents at the start of the growing season - winter–flowering succulents in summer and epiphytic cacti after flowering.

Repotting

All plants must be repotted periodically to allow root growth. Container grown plants need repotting every two or three years to renew their compost. Best time to repot is at the start of the growing season. Choose a clean new pot that is large enough in relation to the overall size of the plant in order to accommodate its new root growth. Most cacti and succulents, except tall-growing plants, are shallow rooted so use pans rather than full-depth pots. Examine the roots and control any pests. Do not overwater after repotting until new roots develop. If the same set of compost is being used as in the old, break up the root ball lightly to help the roots grow into the new compost.

PLANT PROTECTION

Pests

Vine Weevils

They are resistant to most chemicals. The white larvae eat roots and burrow into stems, causing the plant to collapse and die. When repotting, look for tunnels in the compost. If any exist, replace the compost.

Mealy Bugs

An infested plant is tiny, woolly with white nest and sticky honeydew. Treat it either with a systemic or soapy contact insecticide or in temperature over 21°C introduce the predator ladybird *Cryptolaemus montrouzirei.*

Snails and Slugs

These feed on leaves and fleshy stems, leaving unsightly holes and slimy traces. Scatter slug pellets between the plants or apply a preparation based on $Al_2(SO_4)_3$.

Vine Weevils — Mealy Bugs — Snails and Slugs

Red Spider Mites — White Flies — Scale Insects

Fig. 19. Pests

Red Spider Mites

Brown scars are a sign of spider mite attack on young growth. Increase the humidity or use the predator mite *Phytoseiulus persimilis* in temperature over 21°C.

White Flies

These white moth-like insects excrete honeydew on which sooty mould thrive, especially on soft-leafed succulent such as Euphorbia. Use an insecticide spray to treat the plants against these flies.

Scale Insects

These insects check the growth of the plants and the sticky honeydew encourages sooty mould. Control these small, brown limpet-like insects with systemic insecticide.

Diseases

Fungal Leaf Spot

The usually brown or grey spots may merge and cause leaf drop. Tiny fungal spore heads develop where water beads lie on the leaf, leaving sunken brown scars. Burn all infected leaves, treat the plant with copper sulphate, and improve ventilation.

Sooty Mould

Caused by a soil borne fungus, this sooty black mould attacks seedlings that are damaged or weakened by extreme wet conditions. Burn all infected seedlings.

Fungal Leaf Spot

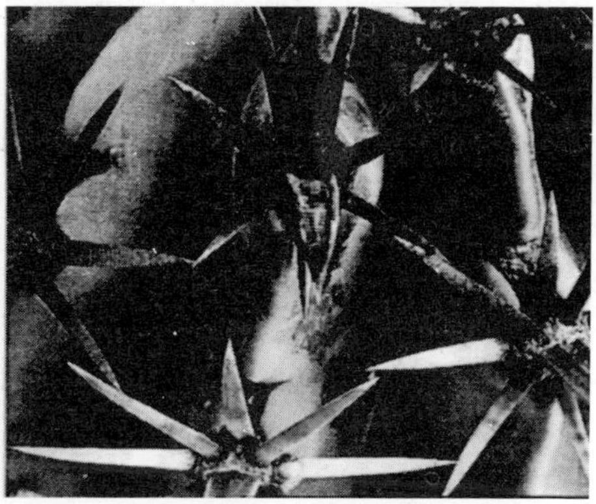

Sooty Mould

Fig. 20. Disease infected plants

CULTURAL DAMAGES

Through natural selection most species of cacti have lost their leaves, which allowed too much evaporation in the desert. Their stems became thick and round to minimize surface area, and to store water. To protect themselves from the Sun and predators, many species developed spines and hair, waxy skin, along with bitter alkaloids.

SUGGESTED READING

1. Bhattacharjee SK 2004 Cacti, in *Landscape gardening and design with plants*, Jaipur: Aavishkar Publishers Distributors, 338-351 pages
2. Bhattacharjee SK 2004 Ornamental succlulents other than cacti, in *Landscape gardening and design with plants*, Jaipur: Aavishkar Publishers Distributors, 351-362 pages
3. Das P and Panda PC 1995 Protected cultivation of cacti and succulents (Volume 12), in *Advances in Horticulture-Ornamental Plants*, [eds] KL

Chadha and SK Bhattacharjee, New Delhi: Malhotra Publishing House, 819-851 pages

4. Dilta BS and Badiyala SD 2004 Cactus has potential in the shivaliks. In *The Tribune* Jan 19
5. Edward F. Anderson. 2001. The Cactus Family, Portland Oregon, Timber Press, 2001, pp 599-611.
6. Hewitt T 1997 The complete book of cacti and succulents, London: Darling Kindersley
7. Kumar R, Singh K and Arora JS 1987 Studies on flower opening in *Mesembryanthemum crinifolium* in relation to light and temperature, *The Punjab Journal* **27** 239-241.
8. Randhawa GS and Mukhopadhyay A 1986 Floricultre in India. New Delhi: Allied Publishers Pvt. Ltd., 274-289 pages
9. Cooper WE and Rochford TC 1973 *Cacti as house plants*, London: Blandford Press
10. Takhtajan 1966 World of succulents, in *Illustrated encyclopaedia of succulents*, Gordon Rowley, Belgium: Leisure Books, 16-17 pages

www.cacti-mall.com

www.cactiguide.com

www.2growsucculents.com

SHRUBS AND CLIMBER

M. R. Dhiman

INTRODUCTION

An ideal home is one in which the house and garden are designed and constructed as one integral part. Today, as never before, the value of fresh air and sunshine is realized and it is obvious that the best way to enjoy these is one's own home garden (Grindal 1963). Gardening is a science because precise knowledge regarding materials, practices, diseases etc. is required to become a really good gardener. It creates around us a world of romance and love, and therefore, of art. The garden constantly reminds us 'never take any plant for granted', just as we should never take any human being or situation for granted. 'To each one according to his need' is the right motto for the garden. A garden in a home is a place of beauty and relaxation and therefore, let the choice of plants, design, harmony and balance be as one would like them and not as any other gardener advises (Trivedi, 1983). The origin of the garden is rooted in the human desire to be surrounded by plants. The great variety of plants permits us to select, in addition to the usefulness, the esthetically pleasing plants. Their pleasant fragrance, as well as their beauty plays an important role in this selection. Plants are the main ingredient of the gardens; they are not static but change seasonally and with time. Plants in the home garden provide food and other life-supporting commodities, and are very important for survival of human beings and other organisms; besides they protect our environment and maintain nature. The evidence of man's dependency on plants for his survival can be demonstrated by palaeo-ethanobotanical findings from prehistoric archaeological sites (Renfrew, 1963). The wild edible plants with high diversity are widely distributed in mountain forests and are valuable source of food and medicines for domestic and commercial purposes. They also have good ornamental value and thus, add beauty to the home gardens.

Shrubs and climbers are as essential part of a garden as are the trees and other ornamentals. They add height, colour, fragrance and texture to any garden, particularly, flowering and foliage types, making great addition to any garden.

ORNAMENTAL SHRUBS

A shrub may be defined as woody, semi-woody or herbaceous perennial plant having branches arising from the base of the plants. A shrub is generally erect and bushy but some are decumbent or prostrate and attaining a height of 0.5-4 m. These are the most important garden plants not because of the large number of cultivated species and varieties but also due to their wide range of variation in shapes and sizes of the plants, leaves, growth habit, and size, shape and colour of flowers, barks and berries. Shrubs are perennial in habit and need little care and grow in all types of soils. Shrubs flower regularly every year and produce flower of almost all the important colours. The shrubs show great variation in heights and thus plants can be chosen from a large variety for growing in various places according to their suitability and utility. In addition to their utility as flowering plants, shrubs are also used for hedges and screening. Hardy dwarf shrubs are usually grown in rockery to add colour and to break the monotony of a large open space or lawn or to separate one part of a garden from the other. The tall shrubs can be used for blotting out some undesirable view. Some of the flowering shrubs help to supply early pollen and nectar that is so necessary for the honey bees (Shewell-Cooper, 1961).

CLASSIFICATION OF SHRUBS

There can be several possible classifications of woody plants. For the botanist, it can possibly be most logical to base the classification on families and genera within those families; but for the horticulturist, classification by their winter effects seems to be the most advantageous. They might also be classified according to their heights, but such a method is better adapted to one section of the country than to the world as a whole, since many species will obviously be more vigorous in warmer climates (Laurie and Ries, 1950). The usual horticultural classification of shrubs is as follows:

1. Shrubs that are grown for their ornamental foliage and beauty
2. Flowering shrubs
3. Shrubs bearing attractive berries etc.

Shrubs can also be classified according to their growth habits i.e. tall, medium and dwarf.

SELECTION OF SHRUBS

The selection of shrubs for a particular location will depend on several factors such as availability of space, sunlight, humidity, temperature, spread and height of shrubs, the type of shrub, i.e. whether flowering, foliage, evergreen or deciduous, fragrant flower, colour of flowers etc. Although selection of individual species and varieties will vary to some extent with sections of the country and individual location, certain definite limitations may be set. Wider growing shrubs such as the Marrow honeysuckle and *Forsythia* are obviously unsuited for planting in confined spaces of a few feet in width. Similarly, shrubs with more upright habits of growth such as *Hibiscus syriacus* and *Philadelphus lemoinei* are not desirable unless used in quantity for screening an undesirable object. *Philadelphus coronaries* or even *Spiraea* var. *honttei* is obviously out of place when planted in front of windows or the average porch, just as the *Deutzia gracilis* or *Spiraea thunbergi* would be of relatively little value for hiding an unsightly shed.

Some shrubs do better in certain regions, and perform poorly or do not survive at all in certain other regions. Plant hardiness is such a variable factor that any recommendations' regarding it is purely relative. Certain areas in northern hills (parts of Himachal Pradesh and Jammu and Kashmir) are covered with snow throughout the winter. Rainfall, prevailing winds, time of fall and spring frosts, soil drainage, and soil moisture are some of the important factors to be consider while selecting the shrubs for hilly areas. Many of the shrubs are grown for their foliage. The size of the foliage, its colour, texture, arrangements on the stem, the time of its appearance in the spring, its condition throughout the season, the time of its fall in autumn and autumn coloration should all be thought of. Extremely coarse-foliaged shrubs such as *Hydrangea quercifolia* are not recommended for large buildings except as foliage accents. On the other hand, the use of small-foliage shrubs such as *Spiraea thunbergi* with large buildings is too great a contrast. Evergreens are another factor in foliage effect. Needled evergreens often contribute to an appearance of somberness and monotony.

Any plant whether in leaf or not, has a definite character. The habit of branching, whether it is strikingly horizontal as in some forms of *Crataegus* and *Viburmum tomentosum* or essentially vertical as in the *Althea*, can be used to advantage. The amount of twigs may be adequate for a screen in winter after the leaves have been shed, as in the case of the privets and most of the shrubbery *Cornus*; or it may not, as in the case of *Hydrangea quercifolia*. The coloured bark of such shrubs such as *Cornus alba sibirica* and the *Kerria japonica* may be used effectively, as may the conspicuous, corky-ridged twigs of *Euonymus alatus*. In addition to seasonal bloom, ornamental fruits contribute greatly to the landscape picture, not only in summer but especially in fall and winter after the leaves have been shed.

PLANT MATERIAL AND PLANTING SEASON

The ideal way to buy the planting material is to go to a reliable nurseryman, preferably somewhere in the local areas during the spring, summer or autumn, and to see the shrubs growing in their nurseries. The other way by which the shrub business may be handled is that you send your order to some small nursery; they pass the order to the wholesaler, sending him labels with the retail nursery's name on it. The wholesaler then digs up the shrubs and sends them to you directly. But the best way to purchase the planting material is from local nurseries situated in the growing areas. It is important to buy the shrub you require under its Latin name. Common names unfortunately differ from area to area and particularly from the North to the South. The Latin name is the international name and the shrub will be known by that name all over the world.

The majority of deciduous shrubs can always be planted during the dormant season when their leaves drop (preferably during November), and the soil is in satisfactory conditions to work. On the other hand, the evergreen shrubs, which keep its leaves in the winter months should always be planted with quite a good ball of soil to its roots and preferably in the spring, i.e. in May when the ground is warm and the weather is showery. The alternative period is from the mid September to mid October. Coniferous evergreens can be planted almost any time except when in active growth during June, since they are balled and burlapped. Adequate soil preparation is not only beneficial but essential for maximum results. Although individual shrubs may be planted in pockets of good soil, complete preparation by the incorporation of organic material is better. Partially decomposed humus in the form of Farm Yard Manure (FYM) can be used. All the plants should be planted at the depth at which they were originally growing. After planting the shrubs, the soil around it should be raised to a height of 15-20 cm to prevent water logging. The spacing in the shrubbery depends upon the size and nature of growth of the plants. When planting singly in rows, the distance will also depend on the growth habit of the shrub, but to some extent this is conditioned by the availability of space and the personal preference of the grower. Large, medium and bushy shrubs are planted at a distance of 1.5- 2.5 m depending on the nature of growth. Shorter plants with less spreading habit can be grown at a distance of 0.5- 1.5 m.

PIT AND MIXTURE PREPARATION

As in the case of large trees or roses, shrubs also need initial preparation of the soil. When planted in shrubbery, the entire bed should be dug up thoroughly. For the larger shrubs, which grow 1.5 m and above, pits measuring 75 x 75 cm should be dug at the places where the shrubs are to be planted. For dwarf shrubs, thorough digging up to a depth of 45 cm will normally be sufficient. The dug-up soil should be mixed thoroughly with well rotten FYM at the rate of 4.5 kg per pit for large shrubs, and 4-5 kg per square metre for other shrubs. It will be beneficial if a handful of bone meal is added to each shrub at the time of planting. Aldrin 5% or B.H.C. 10% @10g per pit should also be mixed with the soil to prevent damage by termites. The digging up of pits or beds should start just at the beginning of the rains, so that the dug-up soil settles down with the first two or three showers.

PROPAGATION TECHNIQUE

Propagation of shrubs is an important aspect. In nature plants multiply themselves by means of seeds and different vegetative parts. In a garden, a gardener has to learn the different methods of propagation by which the plants are multiplied either to replace an old or diseased stock or to meet the increasing demands of an expanding garden. Broadly, the shrubs are propagated by two ways - either sexually (seeds) or asexually (vegetative).

Sexual Propagation

Few of the ornamental shrubs are propagated through seeds. Raising of shrubs from seeds is a simple and economical way to propagate. Collect and clean ripe seeds or buy some and store them in a dry place before sowing. The seeds of some shrubs have a form of dormancy that prevents them to germinate in adverse conditions. Scarification or stratification can brake this dormancy. Sow seeds of hardy shrubs in a cold frame in autumn and in the open for those do not need daily attention. The shrubs, which are propagated from seeds are *Cotoneaster, Holmskiolodia sanguinea*, and *Camellia* etc.

Vegetative Propagation

Most of the shrubs are propagated vegetatively. There are many ways to raise new plants through vegetative means such as taking cuttings, layering, division and grafting. Taking cuttings is a simple way of propagating many shrubs and is useful for cultivars, hybrids and spores. Layering and division are also used for some shrubs. Grafting requires experience and skill compared to other methods but for some shrubs this is the most suitable method.

Propagations by taking cuttings can be done with different type of cuttings:

(i) Soft-wood cuttings: Many ornamental shrubs in the home garden may be propagated by softwood cuttings. Softwood cuttings are taken from the current season's growth from late May to early July. Cutting material should be flexible but mature enough to snap when sharply bent. *Lilac, Forsythia, Weigela, Barberry, Potentilla,* and *Viburnum* are some of the shrubs that may be propagated from softwood cuttings. A proper rooting medium is needed to root softwood cuttings successfully. The rooting medium must not only retain moisture but also drain well and provide physical support. Coarse sand, perlite and vermiculite are good rooting materials. When taking cuttings, plant material should be removed with a sharp knife. Softwood cuttings should approximately be 4-6 inches long. Pinch off the leaves on the lower half of the cutting. Make a fresh cut just below the point where one or two leaves are attached to the stem (node) and then dip the base (cut end) of the cutting in the root-promoting hormone. Insert the cutting approximately 2 inches deep into the rooting medium. Cover the cuttings to reduce water loss and then place the cuttings in bright light, but not direct sunlight. Rooting of most deciduous shrub species should occur in 6-8 weeks.

(ii) Greenwood cuttings - Suitable for deciduous shrubs, and can be of two types:

1. **Nodal cuttings or heel cuttings**: Taken in late spring or early summer.
2. **Semi-ripe cuttings**: Suitable for evergreen and some deciduous shrubs. These can be of two types: mallet cuttings or leaf-bud cuttings. These can be taken from mid to late summer or early autumn. It was found that in *Berberis aristata,* apical cuttings treated with 5000 ppm IBA concentration performed better in sprouting (85%) and rooting percentage (50%) (Ali et al 2008).

(iii) Hardwood cuttings - Suitable for many deciduous and some evergreen shrubs, and are taken from late autumn to mid winter.

(iv) *Root cuttings* – Suitable for some shrubs only, and taken from mid to late winter.

Propagation by Layering

Layering is a method of propagating shrubs whereby a stem is encouraged to develop roots before being removed from the parent plant. There are different types of layering:

- Simple layering: Suitable for many deciduous and evergreen shrubs; done in autumn or spring
- Air layering: Suitable for shrubs with branches that are difficult to lower to the ground level; done in spring. E.g. *Camellia japonica, Magnolia* sps., *Gardenia, Hibiscus sps.,* Hamamelis
- Tip layering: Suitable only for few shrubs; done in spring
- Dropping: Used to raise dwarf shrubs; done in the dormant season
- Stooling: Used for some deciduous shrubs from engrafted plants; done in spring

- Trench layering: Used to increase the stock of deciduous shrubs; done in the dormant season

Propagation by Division

Division is a simple way of increasing shrubs that produce suckers.

Propagation by Grafting

Grafting is one of the most important ways to propagate shrubs that are difficult to root or are unlikely to root from seeds. There are three methods used for this technique:

1. Saddle grafting: Used to propagate evergreen rhododendron specie; done in late winter or very early spring
2. Spliced side-veneer grafting: Used to a large variety of evergreen and deciduous shrubs; done in summer or winter
3. Apical-wedge grafting: Suitable for some species of shrubs; done in mid winter

Training and Pruning

Every shrub, from the smallest flowering plant to the tallest tree, has a definite height at which it attains maximum attractiveness. Since this cannot be changed by artificial treatment, natural growth should be encouraged at all times. This may be accomplished more perfectly by proper pruning than any trick methods. Plants drop their flowers, fruits or foliage at different times in the year. It is not possible to prescribe specific pruning periods for all the species in this limited space,. One general rule, however, applies to all plants - Trim them as soon as they have produced what they were grown for, whether fruit or flower.

Pruning is defined as to remove dead or crossing branches whereas training is to train the plant in a proper shape. Most shrubs can only be pruned at certain times of the year, usually when they are dormant. Shrubs that flower in the spring should only be pruned after the flowers fade, while shrubs that flower later in the summer should only be pruned in early spring. Proper pruning enhances the beauty of almost any landscape shrub, while improper pruning can ruin or greatly reduce its landscape potential. The main objectives of pruning are to:

1 Train the plant

2. Maintain plant health

3 Improve quality of flowers, fruit, foliage or stems

4. Restrict growth

5. Shape plants in an artificial form

Pruning of Deciduous and Flowering Shrubs

Pruning for most deciduous shrubs consist of thinning out, gradual renewal and rejuvenation pruning.

Thinning out involves cutting off a branch or a twig at its point of origin from either the parent stem or ground level. This pruning method results in a more open plant. It does not stimulate excessive new growth, but does allow room for growth of side branches. Considerable growth can be cut off without changing the plant's natural appearance or growth habit. Plants can be maintained at a given height and width for years by thinning out.

In gradual renewal pruning, a few of the oldest and tallest branches are removed at or slightly above ground level on an annual basis (Fig. 21). Some thinning may be necessary to shorten long branches or maintain a symmetrical shape. To rejuvenate an old, overgrown shrub, remove one-third of the oldest, tallest branches at or slightly above ground level before new growth starts.

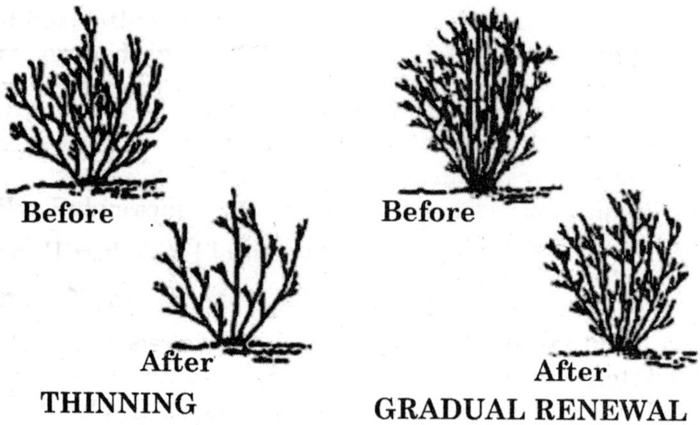

Fig. 21. Deciduous shrub pruning

If a shrub is grown for its flowers, timely pruning is essential to minimize the disruption of blooming. Spring flowering shrubs bloom on last season's growth and should be pruned soon after they bloom. This allows for vigorous summertime growth and results in plenty of flower buds the following year. Some examples of shrubs that bloom on last season's growth are Japanese quince, All forsythia, Honeysuckle, Hawthorne, Azaleas and Rhododendrons, Shrub roses, White spireas and Viburnums. Some shrubs that bloom after June usually do so from spurs, which are formed on shoots that grow the same spring. These shrubs should be pruned in later winter to promote vigorous shoot growth in spring. Examples of such shrubs that bloom on current season's growth include Butterfly bush, Shrub althea, Hydrangea, St. Johnswort, Crape myrtle, Bush rose, flowering almond and Viburnum.

Pruning Evergreens and Narrow-leafed Shrubs

Since narrow-leafed evergreens produce new growth in spring and fall, and do not grow much in summer, they should be pruned in the first or second week of May or June. Evergreens should be pruned according to their growth habits. These plants should be allowed to assume their natural shape. Pruning is a matter of cutting the branches so that a more desirable plant is attained through compact, controlled growth. This requires pruning individual stems rather than shearing. Shearing not only ruins the natural growth habit but prevents light from penetrating into the center of the plant resulting in foliage drop. Broad-leafed evergreens such as Gardenias, Camellias, Azaleas, Pyracantha, Hollies and Photenias require very little pruning. Lightly thin, broad-leafed evergreens grown for their showy fruit such as pyracantha and holly should be pruned during the dormant season. Old or weak stems should be removed. This group can go several years without pruning except for some slight cosmetic pruning to keep them neat.

Pruning and Training of Hedges

Hedges are a row of plants that merge into a solid linear mass. They have served gardeners for centuries as screens, fences, walls and edges. A well-shaped hedge is no accident. It must be trained from the beginning. Establishing a deciduous hedge begins with the selection of nursery stock. Choose young shrubs 1-2 feet high, preferably multiple-stemmed.

When planting, cut the plants back to 6 or 8 inches;as this induces low branching. Late in the first season or before bud-break in the next season, prune off half of the new growth. The following year, trim off the half again. In the third year, start shaping. Trim to the desired shape before the hedge grows to its desired size. Never allow plants to grow untrimmed to the final height before shearing. Do not allow lower branches to be shaded out. After the hedge has reached the desired dimensions, trim closely in order to keep the hedge within chosen bounds.

Table 6. Spring-flowering shrubs that should be pruned after flowering

Scientific name	Common Name
Berberis	Barberry
Chaenomeles	Flowering quince
Cotoneaster	Cotoneaster
Crataegus	Hawthorn
Deutzia	Deutzia
Euonymus	Winged spindle tree
Forsythia	Forsythia
Kalmia latifolia	Mountain laurel

Scientific name	Common Name
Lonicera	Honeysuckle
Magnolia	Magnolia
Malus	Crabapple
Philadelphus	Mock orange
Prunus	Flowering cherry and plum
Pyracantha	Firethorn
Rhododendron	Rhododendron and Azalea
Rosa	Shrub roses
Spiraea thunbergii	Thunberg spirea
Spiraea x vanhouttei	Vanhoutte spirea
Styrax japonica	Japanese snowball
Syringa	Lilac
Viburnum	Viburnum
Weigela	Weigela

Evergreen hedging need not be as small as deciduous material and should not be cut back when planted. They should be trimmed lightly after a year or two. Start shaping as the individual plants merge into a continuous hedge. Do not trim too closely because many needle-bearing evergreens do not easily generate new growth from old wood. Hedges are often shaped with flat tops and vertical sides; however, this unnatural shape is seldom successful. As far as the plant is concerned, the best shape is a natural form, with a rounded or slightly pointed top and with sides slanting to a wide base (Fig. 22 and Fig. 23).

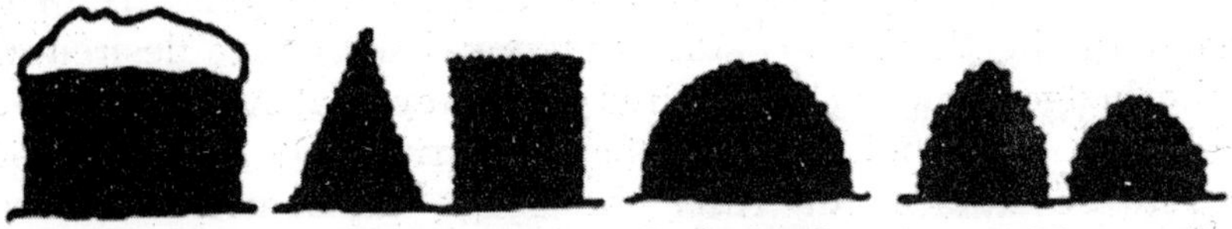

Fig. 22. Hedge Shapes

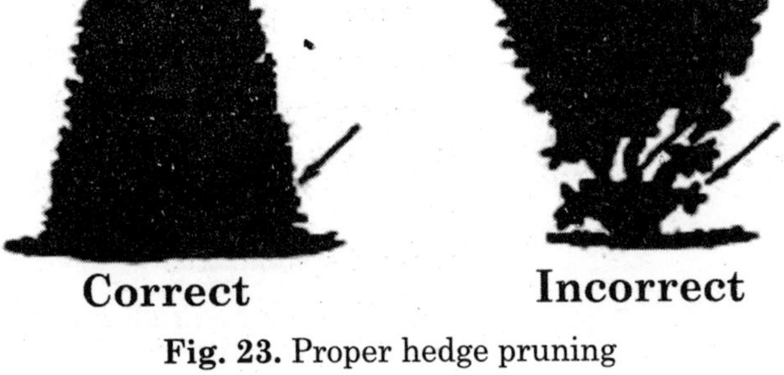

Fig. 23. Proper hedge pruning

Table 7. Summer-flowering shrubs that should be pruned before spring growth begins

Scientific name	Common Name
Abelia x grandiflora	Glossy abelia
Buddleia davidii	Butterfly bush
Hibiscus syriacus	Shrub-althea
Hydrangea arborescens	Smooth hydrangea
Hydrangea paniculata	Pee Gee hydrangea 'Grandiflora'
Hypericum	St. Johnswort
Magnolia virginiana	Sweet bay
Rosa cvs.	Hybrid tea roses
Sorbaria	False-spirea

Table 8. Shrubs that may be pruned both before and after bloom

Scientific Name	Common Name
Cornus sericea	Red-osier dogwood
Cotoneaster apiculatus	Cranberry cotoneaster
Cotoneaster divaricatus	Spreading cotoneaster
Cotoneaster multiflorus	Many-flowered cotoneaster
Mahonia aquifolium	Oregon hollygrape
Spiraea x bumalda	Anthony Waterer and Froebel spirea
Weigela	Weigela

Table 9. Shrubs that may be rejuvenated when completely pruned (completely cut back to the ground)

Scientific name	Common Name
Buddleia davidii	Orange-eye butterfly bush
Forsythia	Forsythia
Hibiscus syriacus	Shrub-althea
Hydrangea arborescens 'Grandiflora'	Hills-of-snow
Hydrangea quercifolia	Oakleaf hydrangea
Ligustrum vulgare	Privet
Lonicera	Honeysuckle
Spiraea	Spirea
Syringa	Lilac

Hedges of slow-growing plants such as boxwood need to be trimmed sooner. Excessive untrimmed growth will kill lower leaves and will also pull the hedge out of shape. Trimming frequency depends on the kind of shrub, the season and desired neatness. Rejuvenating evergreen hedges is more difficult. As a rule, evergreen hedges cannot stand severe pruning.

DESCRIPTION OF SHRUBS

Flowering and Evergreen Shrubs

Agapetes serpens

Family: ***Ericaceae***

Native: Himalayas (Nepal, Northeast India particularly Darjeeling)

Common Name: Himalayan Lantern

Himalayan Lantern is a beautiful shrub, 40–60 cm tall, with arching stems and tiny hanging lantern-like flowers. Rootstock is tuberous. Twigs are cylindrical and 1–2 mm in diameter. Leaves are crowded, on short stalks of 1 mm. Leaf blade is, narrowly ovate or ovate-oblong, 1.2–2 cm long, 5–8 mm wide, leathery, dark green on the upper side and pale green on the lower side. Base is rounded; margin slightly recurved, toothed above the middle part; tip is sharp or blunt. Flowers hang down in fascicles of 1–3. Stalks are 0.7–2.5 cm, slightly thicker at apex. Sepal cup is 3.5–4 mm, 5-winged. Sepals are ovate-triangular, 3.5–4.5 mm long. Flowers are tubular, 1.2–2.8 cm, 5-angled, and bright red, orange, or pinkish white in colour with dark red zig-zag bands. Tiny petals are curved back, triangular, 2–3 mm long. Berry is obovoid, 5-winged, about 6 mm in diameter. Persistent sepals are enlarged. Flowering season for these shrubs is from May–June. These are easily propagated through cutting.

Azaleas

Family: ***Ericaceae***

Origin: China, Japan

Common Name: Azalea

Botanically, the azaleas are included in the genus Rhododendron. Some important species are *R. indicum, R. kaempferi, R. ledifolium* and *R. poukhanense*. Some other hardy deciduous azaleas are *R. arborescens, R. canadense, R. japonicum, R. nudiflorum, R. occidentale* and *R. rhombicum*. Azaleas do not like lime soils. They need to be cultivated carefully and should be grown in sheltered position. A light soil with plenty of leaf mould is good for azaleas. The flowering season for these shrubs is during June-August and the flowers are either single or double and come in white, pink, rose, scarlet red and purple colours. Azaleas are multiplied by divisions, layering or seeds. Azalea leafy gall can particularly be destructive to azalea leaves during the early spring. Hand-picking infected leaves is the recommended method of control

Camellia japonica

Family: ***Theaceae***

Origin: Japan, China

Common Name: Japanese camellia

This is a large evergreen shrub or small tree, with shining evergreen elliptic-lanceolate pointed leaves, 2-4 cm across during October to November. The flowers are erect, white in colour, almost stalk less. Species *C. japonica* bears pinkish-red double flower during February-April. It is propagated through semi-hardwood cutting, air layering or by seeds. Camellia is infected by some fungal and algal diseases including Spot Disease, which gives the upper side of leaves a silver color and round spots, and can cause loss of leaves, Black Mold, Leaf Spot, Leaf Gall, Flower Blight, which causes flowers to become brown and fall, Root Rot, and Canker.

Caryopteris incana (Syn. *C. mastacanthus, C. sinensis*)

Family: Verbenaceae:

Native: Eastern and Southern Asia

Common Name: Sunshine blue

The shrub grows to a height of 1.8-2.5 m and profusely produces lavender blue flowers during winter and spring. The shoots are pruned to 50-70 cm during the rains. Propagation is by cutting.

Hamamelis mollis

Family: ***Hamamelidaecae***

Native: Central and Eastern China

Common Name: Chinese Witch-hazel

Hamamelis (Witch hazel) species and varieties are large shrubs or small trees that bring flower and leaf gold to temperate landscapes when gardeners least expect color and fragrance. Their diversity of bloom times and leaf colours, mature shapes, ease of maintenance, and lack of serious disease and insect-pests provide an extensive woody plant collection that can strengthen landscape designs. It bears blue, red or yellow coloured flowers in late winter during January-March. Flowers are set close to the bare branches. The leaves turn bright yellow in late fall. They grow well in light shade. It is propagated by layering and grafting on *H. virginiana*.

Hibiscus spp.

Family: ***Malvaceae***

Native: East Asia

Common Name: China rose and shoe flower

This is a group of flowering shrubs consisting of several species and hybrids. The important temperate species include *H. rosa-sinensis*, a large, 1-2 m tall and bushy shrub having numerous branches. The broadly ovate, sharp-toothed green leaves are ornamental. The fine-petalled large, single or double, rose-scarlet coloured flowers are borne terminally on the branches. It is propagated by air layering, cuttings or by budding.

Holmskioldia sanguinea

Family: ***Verbenaceae***

Native: Himalayan lowlands, China

Common Name: Cup and saucer, Chinese hat, parasol flowers.

This is a straggling climbing shrub with broadly ovate leaves and scarlet tubular flowers with large conspicuous red cup-shaped calyx, borne in crowded axillary clusters during August-December. It requires full to partial sunlight. It is easily propagated by semi-ripe cuttings in mid to late summer.

Inula cappa

Family: ***Compositae***

Origin: Nepal

Common Name: Dware Phool (In Nepal)

This is an aromatic shrub with woolly or silky-haired branches, leaves and inflorescence, with oblong-lanceolate leaves, and with numerous small yellow flower heads in large terminal domed clusters of 8 cm or more across. These appear during September-October. These are distributed throughout Nepal to about 2500 m on open, sunny hillsides; also in Northern India, Bhutan, China, Thailand, Malaysia, and Java. The propagation is from seeds.

Kalmia latifolia

Family: ***Ericaceae***

Native: Eastern United States

Common Name: Mountain–laurel or spoonwood

It is a broad-leafed, evergreen shrub, 3-6 m tall. Its flowers are round, ranging from light pink to white colour, and occurring in clusters from May to June. Mountain-laurel is poisonous to several animals including horses, goats, cattle, sheep, and deer due to presence of andromedotoxin and arbutin in it. It is not toxic to dogs, cats, or small household pets. The green parts of the plant, flowers, twigs, and pollen are all toxic. It is propagated from seed, cutting and tissue culture.

Lonicera rupicola

Family: ***Caprifoliaceae***

Origin: Himalayas

Common Name: Bush honeysuckle

This is a dense low growing shrub with stiff interlacing branches, which become somewhat spiny, with tiny oblong leaves. Flowers are almost stalk-less and in clusters of scented, pale-pink to purple coloured flowers, which appear during June-July. It is propagated through cuttings.

Mahonia napaulensis

Family: ***Berberidaceae***

Native: Nepal, Northeast India

Common Name: Nepal Mahonia, Indian barberry

Indian barberry is an evergreen shrub that grows to 2.5m to 3m, with large, pinnately compound leaves. The leaves are about 18 inch (46 cm) long with 9-13 stiff, sharply spiny, holly like leaflets. The fragrant lemon-yellow flowers, appearing in late winter, are borne in erect racemes 3-6 inch (7.6-15 cm) long. The fruit is a berry that is first green in colour, then turns bluish black with a grayish bloom. The fruit is about half inch long and hang in grapelike clusters. It is in leaf all year, and in flower from March to April. The fruit is eaten raw or cooked. The fruit can also be dried and used as raisins. It is propagated through cuttings.

Osbeckia nepalensis Syn. *O. speciosa*

Family: ***Melastomaceae***

Native: Nepal

Common Name: Nepal Osbeckia

This is a dwarf (45 cm), much-branched shrub, having lanceolate leaves, which are often spotted brown. The purplish rose flowers appear during June. It is propagated through cuttings or seeds. Other species is *O. stellata*, a shrub often hanging from rocks and banks, with large pinkish-purple flowers in terminal clusters that appear during July-November.

***Paeonia saffruticosa* (Syn. *P. montan, P. arborea*)**

Family: ***Ranunculaceae* or *Paeoniaceae***

Native: China, Japan

Common Name: Tree peony, Mountain peony

This shrub is about 2 m tall. The flowers have a close resemblance to rose and may be single, semi-double or double. The flowers vary in colour from white to rose-pink and red flowers during May to June. The plant thrives best at medium to high altitudes. It should be planted in sunny locations of the garden. It is propagated by air-layering or grafting.

Potentilla polyphylla

Family: ***Rosaceae***

Native: Himalayas, Northeast India

Common Name: Many Leaved cinquefoil

This shrub is distinguished by its leaves, which have many elliptic toothed leaflets, the larger 1.3-4 cm, alternating with much smaller more rounded leaflets. Flowers are rather small (8-15 mm across), yellow in colour, and appear during June-July. It is propagated by layering or softwood cuttings or by seeds. Other species includes *P. fruticosa, P. coriandrifolia,* and *P. atrosanguinea.*

Pyracantha angustifolia

Family: ***Rosaceae***

Native: China

Common Name: Narrowleaf firethorn

This is an evergreen, thorny, protective and large shrub suitable for town gardens. They like to be planted in shade or semi-shade. They have white or yellow blossom in June, but these are seen insignificant compared to the bright orange or red berries that appear during winter. It can be used to make hedges for home security. It is propagated from seeds and cuttings.

Reinwardtia indica

Family: ***Linaceae***

Native: North India

Common Name: Yellow Flax or Pyoli, Basanti

Yellow Flax is an erect to spreading shrub, up to 1 m tall, and is commonly found only in a prostrate state. Leaves are elliptic to inverted lance-shaped. On a midday in December or January, you may witness as many as two dozen yellow flowers that are open. Each golden-yellow flower, 2 inch wide, lasts for one day and is described as being fugacious, meaning that the corolla withers and falls off easily. The corolla is composed of five petals fused to form the 2 cm tube. The golden-yellow colour of the corolla is improved by the presence of fine reddish veins; such lines are termed nectar guides or nectar lines, because they typically communicate with the pollinating insects where to go to find a nectar reward. The five corolla lobes are cleverly overlapped (imbricate). Emerging from the floral tube are observed three curiously shaped green stigmas on three styles nearly 2 cm in length. Yellow Flax is found from Pakistan to Southwest China, at altitudes up to 1800 m. It is a common wildflower of North-Indian hill stations. The flowering season for this plant is during November-May. It is propagated through semi-hardwood cutting or seeds.

Sarcococca coriacea

Family: ***Buxaceae***

Origin: Himalayas, Nepal

Common Name: Phitiphiya (Nepali)

This is a small evergreen shrub up to 1 m, with green stems, and shinning, leathery, lanceolate leaves. Flowers are white in colour, and in short dense axillary clusters of 6-10 mm. The flowering season is from September to May. The propagation is by semi-hardwood cuttings.

Skimmia anquetilia

Family: ***Rutaceae***

Native: Himalayas

Common Name: Skimmia

This is a strongly aromatic evergreen shrub or small tree, and creeping or erect up to 150 cm tall. It bears very compact terminal clusters of yellow flowers during April-May. The skimmias have a rather slow growth and they thrive best in a sandy loamy soil, but also grow well in strong clay. They prefer a partly shaded condition. On account of their handsome fruits they are sometimes cultivated in pots in a sandy compost of peat and loam. It is propagated from cuttings or seeds.

Tibouchina elegans (Syn. *Pleroma elegans*)

Family: ***Melastomaceae***

Native: Brazil

Common Name: Glory Bush

This is a 150 cm tall shrub. The flowers are small, 3.6 cm across, solitary, purple in colour and produced in abundance during May and continue up to October. It is propagated through semi-hardwood cuttings. The other ornamental species include *T. semidecandra* (**Syn**. *Lasiandra macrantha*). The flowers of this species are rich purple in colour, very large, 8-13 cm across in size, and are borne solitary or in clusters of a few flowers throughout the year in a glasshouse but during the autumn in theoutdoors.

Vaccinium corymbosum

Family: ***Ericaceae***

Native: Eastern North America, Europe and Asia

Common Name: Blueberry

This is a tall (3 to 4 m) deciduous shrub, worthy for its white-tinted flowers and blue-black globose berries which are edible. It flowers during March-April. The leaves turn rich-red in the autumn. This plant grows best in wooded or open areas with moist acidic soils. It is propagated by layering or cuttings.

Vaccinium vacciniaceum

Family: ***Ericaceae***

Native: East Asia to Eastern Himalayas

Common Name: Blue berry, Cranberry

This is an evergreen shrub. The flowers are greenish, pale yellow, white or pinkish in colour, and borne in one or several stalked one sided clusters at the ends of the leafy branches. It flowers during April-June. It can grow in semi-shaded areas or areas with no shade. It is propagated from seeds, layering in late summer or early autumn or division of suckers in spring or early autumn.

Foliage and Flowering Shrubs

Aucuba japonica

Family: ***Cornaceae***

Native: Eastern Asia and Japan

Common Name: Spotted laurel

Aucuba japonica or Gold Dust Aucuba is an attractive broad-leafed evergreen shrub. They are slow to moderate growers and reach a height of 4-12 ft (1.2-3.7 m) with a spread of 4-6 ft (1.2-1.8 m). Leaves are opposite, elliptical to elliptical-lanceolate, 3-7 inches (8-18 cm) long and 1-3 inches (2.5-7.6 cm) wide. Leaf colours are dark green with golden yellow variegation. There are many cultivars with different leaf variegation patterns. They are the best evergreen for deep shady conditions and are very drought resistant. Flowers are small and inconspicuous. These shrubs are planted in deep shade underneath the trees. It is very easily propagated from cuttings.

Euonymus japonica

Family: ***Celastraceae***

Native: Japan, Korea, China

Common Name: Japanese Spindle

This is a group of shrubs which are not grown for beauty of their flowers, but many of them are extremely fine in fruit and develop good colouring. The flowers are inconspicuous, greenish-white in colour, and 5 mm diameter. In the fall season, the orange fruit hangs below the flaring pink seed coverings. It needs to be pruned regularly around the year. It is easily propagated form semi-hardwood or hardwood cuttings during summer or fall.

Juniperus communis

Family: ***Cupressaceae***

Native: Temperate Northern hemisphere

Common Name: Jupiter berry

This is a dense, erect, evergreen, spreading shrub, which grows up to 1.5 m. It has needle like sharp pointed leaves, 6-13 mm long, with a broad bluish white band above, in whorls of three. The seed cones are berry-like, green in colour, ripening in 18 months to purple-black with a blue waxy coating. The fruits are spherical, 4–12 mm in diameter, and usually have three (occasionally six) fused scales, each scale with a single seed. Juniper berries have long been used as medicine by many cultures (Sati and Joshi, 2010). It is propagated through seeds. Other important species include *J. recurva, J. squamata, J. indica* and *J. macropoda*.

Sambucus nigra

Family: ***Caprifoliaceae***

Native: Europe including Britain

Common Name: European Elder, Elderberry

This is a deciduous shrub or small tree, 4–6 m (rarely to 10 m) tall. The bark, light grey when young, changes to a coarse grey outer bark with lengthwise furrowing. The leaves are arranged in opposite pairs, 10–30 cm long, pinnate with five to seven (rarely nine) leaflets. The leaflets are 5–12 cm long and 3–5 cm broad, with a serrated margin. The flowers are borne in large corymbs of 10–25 cm diameter in mid summer. The individual flowers are 5–6 mm in diameter, with five petals and white in colour. The fruit is a dark purple to black berry, 3–5 mm in diameter, produced in drooping clusters in the late autumn. They are an important food for many fruit-eating birds. It is propagated from seeds and hard-greenwood cuttings.

Buxus wallichiana

Family: ***Buxaceae***

Native: Western and Southern Europe, Southwest, Southern and Eastern Asia, Africa, Madagascar, Northernmost South America, Central America, Mexico

Common Name: Himalayan boxwood

They are slow-growing evergreen shrubs and small trees, growing to 2–12 m (rarely 15 m) tall. They are commonly found in Western and Central Himalayas at an elevation ranging from 1200 m to 2900 m, where it grows gregariously in patches in moist shady ravines. Extracts of the leathery leaves and bark are used in local medicine (Govt. of India 1986). The leaves are opposite, rounded to lanceolate, and leathery. They are small in most species, typically 1.5–5 cm long and 0.3-2.5 cm broad. The flowers are small and yellow-green in colour. Box are commonly grown as hedges and for topiary. It is propagated by seeds or semi-hardwood cuttings.

Fragrant Flowering Shrubs

Abelia x grandiflora (Syn. *A. chinensis*)

Family: ***Caprifoliaceae***

Native: Afghanistan to Central Nepal

Common Name: Glossy abelia

It is a deciduous or semi-evergreen multi-stemmed shrub that grows to a height of 1.5-2 m. The pinkish-white, bell-shaped, scented flowers are borne in clusters during summer. The persistent brown sepals are attractive. It prefers semi-shade and well drained soil. The plant is propagated from cuttings.

Aglaia odorata

Family: ***Meliaceae***

Origin: China

Common Name: Chinese perfume plant

This is ashrub about 2 m tall, bearing small, yellow, fragrant flowers during the summer and the rains. It is suitable for shrubbery and as a hedge. It is propagated through cuttings.

Buddleia asiatica

Family: ***Loganiaceae***

Native: Asia

Common Name: White butterfly bush, winter flowering lilac

The sweet scented flowers of this shrub are produced in profusion in drooping terminal or axillary panicles in late winter or early spring. *B. asiatica* blooms on year old wood so it does not need pruning until late spring. It is easily propagated from cuttings. Other important species include *B. davidii, B. lindleyana, B. officinalis* etc.

Clerodendron spp.

Family: ***Verbenaceae***

Native: China

Common Name: Glory bower

C. fragrans is a hardy shrub of 1-2 m height, which grows best in moist and semi-shaded conditions. The shrub bears usually double, white, highly fragrant flowers in large terminal clusters during the summer and rainy seasons. It is propagated from root suckers or cuttings.

Daphne cannabina

Family: ***Thymelaeaceae***

Native: Himalayas, Nepal

Common Name: Lokta bush

This is an evergreen erect shrub, 2-2.5 m tall, which grows naturally in the Himalayas at the altitudes of 6500 to 9500 ft. In the heart of the Himalayas, Nepali craftsmen have been producing Nepali handmade paper from this plant for over a thousand years. The people of Nepal still continue to use Nepali handmade paper or lokta paper in their daily life, and until the Chinese occupation in 1959, they traded it with Tibet where it was used in the monasteries for sacred texts. The shrub bears fragrant purplish-pink flowers from January. It is propagated from cuttings and by layering. Other species include *D. surveil*, an erect deciduous shrub, 2-3.5 m tall. The young branches are densely white-downy. The green palling to white, slightly fragrant flowers appear from January.

Datura suaveolens

Family: ***Solanaceae***

Native: Tropical America

Common Name: Angel's tears, White angel's trumpet, Datura

There are several varieties of Datura, some of which become small trees and bear very large trumpet-shaped flowers that hang down like bells. This particular species is naturalized in most parts of the world. This is a large shrub with large leaves. It has very large, pendulous, funnel-shaped white flowers of up to 30 cm long that are sweet-scented at night. In India, it is often heard of in the Sanskrit scriptures as being one of the favoured plants of the travelling sadhus, particularly worshippers of Lord Shiva, the god of dissolution. With the proper preparation, it is said to have hallucinogenic properties. However, a wrongly prepared dose can be both toxic and fatal. Commonly, Datura species are grown for their attractive flowers. However, all parts of the plant are poisonous. It flowers during May-November. It is propagated by seeds and softwood cuttings.

Deutzia corymbosa

Family: ***Saxifragaceae***

Native: Eastern and Central Asia

Common Name: Not known

This is a vigorous, tall (3 m) shrub, suitable for medium to high elevations. The pure white, small, fragrant flowers appear in dense panicle or corymbs during summer. It is easily propagated from cuttings.

Gardenia jasminoides

Family: ***Rubiaceae***

Native: Asia

Common Name: Cape jasmine

This is a very popular shrub and grown for its highly fragrant flowers. The large, white, double, highly fragrant flowers cover the plants during summer. It is propagated from cuttings or by air layering.

Luculia gratissima

Family: ***Rubiaceae***

Native: Himalayas, Nepal, Tibet to Assam

Common Name: Pleasant Luculia, Ban Kangiyo (In Nepal)

Pleasant Luculia is one of the most attractive autumn flowering shrubs of the Himalayas. It is an evergreen shrub that grows to 3-5 cm tall. It bears large pink, fragrant flowers in large terminal domed clusters, 8-15 cm across. Flowers are 2-2.5 cm across, with slender flower-tube 2.5-4 cm long, opening into 5 rounded, spread-out petals. Sepals are hairy, and they fall off soon. Leaves are entire, variable, elliptic to lance-shaped, long-pointed, and 13-18 cm long. Pleasant Luculia is found in the Himalayas, from Central Nepal to Northeast India and Southwest China, at the altitudes of 1200-1800 m. The flowering season is from September to October. It is propagated from cuttings.

Magnolia stellata

Family: ***Magnoliaceae***

Native: Japan

Common Name: Star magnolia

This is a deciduous shrub that grows 10 ft tall, and flowers during March-April. The flowers are star-shaped, white, purple or creamy in colour, 2-3 inches across, fragrant, many petalled, and freely produced before the leaves appear. It flowers whilst only quite young. The leaves open in bronze-green colour, turning to deep green as they mature, and yellow before dropping in autumn. *Magnolia Stellata* grows best in neutral or slightly acid soil, well enriched with humus, and in Sun or semi-shade. It is propagated by layering in spring or by seeds sown in February.

Nerium indicum

Family: ***Apocynaceae***

Native: India, China

Common Name: Kaner

This is an evergreen shrub, which bears large, terminal branched clusters of mild fragrant red, pink or white funnel-shaped flowers during April-May. It is found all over India especially in Kashmir. Its root and bark have medicinal properties. Its bark, leaf, sap and seeds are toxic. Despite its toxicity oleander (also called Indian oleander) is an effective snakebite cure. It is propagated easily by cuttings.

***Olea fragrans* (Syn. *Osmanthus fragrans*)**

Family: ***Oleaceae***

Native: China, and Japan

Common Name: Sweet olive

This is adelicate shrub or small tree that grows up to 9 m tall. It bears small, sweet-scented creamy-white coloured flowers several times in a year but mainly during June-August. It is propagated by layering.

Osmanthus suavis

Family: ***Oleacae***

Native: India, Bhutan, Nepal and Myanmar

Common Name: Osmanthus

This is an evergreen shrub or small tree that grows up to 7 m, with leathery narrow-elliptic acute, finely toothed or small leaves. They bear short axillary clusters of small scented white tubular 4-lobed flowers that appear during early spring. The propagation is by tip cuttings.

Philadelphus coronarius (Syn. *P. pallidus, P. coronarius nivalis*)

Family: ***Saxifragaceae or Hydrangeaceae***

Native: Southeast Europe, Asia Minor

Common Name: Mock-orange, Syringa

It is a hardy, tall, deciduous, spreading shrub, densely covered with leaves and flowers. The flowers are creamy white, strongly scented and are borne during the spring and early summer. Other species are *P.x lemoinei* and *P. pubescens* (Syn. *P. latifolius*). This shrub is 3-5 m tall. The *Philadelphus* species are suitable for growing only at medium and high elevations, either in ground or in pots. They prefer light shade. The shrubs are pruned just after flowering. These are propagated from cuttings or suckers or layers.

Rhododendron spp.

Family: *Rubiaceae*

Native: Himalayas, Florida

Common Name: Rhododendron

Important species include *R. hookeri*, which bears blood red flowers; *R.keleticum*, which bears purplish-crimson flowers, suitable for rock gardens; *R. Lindleyi*, which bears white, fragrant flowers; *R. schlippen-bachii*, which has pale pink to soft rose colouredspots with red-brown flowers; *R. thomsonii,* which has blood red flowers. These are propagated through greenwood cuttings rooted in sand.

Syringa vulgaris

Family: ***Oleaceae***

Origin: Persia to China

Common Name: Persian Lilac

It is known as lilac and a most popular shrub of the temperate region. The flowers and foliage are most attractive. The flowers are lilac in colour, fragrant and borne profusely in dense terminal spikes during spring. The heart-shaped leaves are attractive. It is propagated by grafting, seeds, semi-hardwood cuttings and layering.

Viburnum grandiflorum

Family: ***Caprifoliaceae***

Origin: Temperate Northern hemisphere

This is a large shrub that grows 3 m tall, and is distinguished by its dense nearly stalkless clusters of relatively large pink fragrant flowers. These flowers appear on bare stems before the leaves during April-May. The fruits are oval, pink in colour, but turn blue-black on maturity. It requires full Sun or partial shade. It is propagated through cuttings. Other important species are *V. cylindricum, V. erubescens, V. cotinifolium, V. nervosum, V. opulus* and *V. mullaha* etc.

Deciduous Shrubs

***Abutilon* spp.**

Family: ***Malvaceae***

Native: Brazil

Common Name: Flowering Mapple

Abutilon grows better at medium to high elevations and attains a height of 1-1.5 m. It propagates from seeds or cuttings. They prefer partial shade and shelter from wind and excessive rains. The flowers are hibiscus-like. In *A. striatum* var. *thompsonii*, the leaves are richly mottled yellow and green. It is grown for its foliage. *A. pictum* is a short shrub with orange-yellow veined crimson flowers, appearing during spring to summer.

Berberis asiatica

Family: ***Berberidaceae***

Native: East Asia to the Himalayas (Nepal)

Common Name: Barberry

This is a yellow flowering shrub, 2-3 m tall bearing purple berry during November. Other species are *B. bhutanensis, B.vulgaris* etc. It is propagated through seeds or cuttings.

***Chaenomeles japonica* (Syn. *Cydonia japonica, Pyrus japonica*)**

Family: ***Rosaceae***

Native: Eastern Asia and Japan

Common Name: Japanese quince or flowering quince

This is a spreading, deciduous shrub that grows 1-3 m high and 6 m wide. This plant is fully hardy and can withstand frost to about -25°C. It produces apple-like, edible, aromatic yellow to green or purplish green fruits in the autumn. They can also be useful as ground cover or low hedging. Scarlet red flowers are borne in 2-6 clusters during March-April. It is propagated through seeds or semi-ripe cuttings of half-ripe wood or from layering.

Contoneaster frigida

Family: ***Rosaceae***

Origin: Himalayas

Common Name: Rockspray

This is a large, deciduous, spreading shrub grown for their flowers and berries. The numerous white flowers are borne on clusters, followed by abundant scarlet fruits in June. It is one of the most beautiful species when in flowering and fruiting. Propagated by seeds or divisions of dwarf creeping type. It is useful for grouping here and there. However, it is valuedfor the beauty of its richly-coloured fruits in autumn. Other ornamental species are *C. horizontales, C. microphylla* etc.

Cytisus scoparius

Family: ***Leguminosae***

Native: Western and Central Europe

Common Name: Common Broom

This is a deciduous, erect, perennial leguminous shrub, growing up to a height of 2.5 m. The bright shining yellow flowers are solitary and appear in late summer. This is one of the hardiest species, which tolerates a temperature of -25°C. It contains a toxic compound that depresses the heart and the nervous system. Generally, it is propogated from seeds but also from cuttings.

Forsythia spp.

Family: *Oleaceae*

Native: China

Common Name: Weeping Forsythia

This is a group of shrubs, which can grow in any soil provided it is well drained. The shrubs are cut back after flowering is over in spring (March-April). Commonly, it is propagated through green-wood cuttings. It requires full Sun and dappled shade. It is best to be used as a specimen shrub. Periodic pruning of old, nonproductive stems at ground level helps to maintain a healthy, vigorous plant by stimulating the growth of the new stems. It should be pruned soon after blooming because the buds for next flowering season are produced on stems that grow the previous summer. Some ornamental species includes *F. x intermedia* var. *densiflora*, which have the flowers that are pale yellow and rather crowded; *F. x intermedia* var. *spectabilis* (*F. spectabilis*), which have flowers that are bright yellow in colour and crowded. This is a beautiful shrub, which flowers in a leafless condition during February-April.

Hibiscus syricus

Family: ***Malvaceae***

Native: China and India

Common Name: Rose of Sharon

It is a deciduous shrub. The colour of flowers vary from white, light pink or red to purple or blue. *Hibiscus syriacus* is the national flower of South Korea. Flowering time is often from the end of July till the first part of October. It is best to use in shrub border, screens, grouping and mass plantings. It is also useful for late season blooming. The species *H. paramutabilis* has large and vigorous flowers. These are easily propagated either from seeds, layering or cuttings. Important cultivars include 'Diana', 'Lady Stanley', 'Ardens', 'Lucy', and 'Blushing Bride'.

Hydrangea paniculata

Family: ***Saxifragaeceae***

Native: China, Korea and Japan

Common Name: Panicled Hydrangea

This is a group of deciduous shrubs, which can be cultivated at medium and high altitudes. The leaf is handsome, large, oval to roundish in shape and toothed at margins except in one species. The flowers are borne in large (20-30 cm in diameter) round heads of white, pink or blue colours during summer. It is propagated from seeds or cuttings.

Hypericum hookerianum

Family: ***Guttiferae***

Origin: South India, Sikkim, Himalayas

Common Name: St. John's wort

This is a small branched deciduous shrub that grows up to 1.5 m height. It bears a cluster of few yellow flowers, each 3-4 cm across, during September-October. It is propagated through cuttings and seeds. Other temperate ornamental species are *H. oblongifolium, H. choisianum,* and *H. uralum*.

***Lagerstroemia indica* (Syn. *L. chinensis*)**

Family: *Lythraceae*

Origin: China, but widely cultivated in India

Common Name: Crape Myrtle, Pride of India

This is a pretty deciduous shrub or small tree, which never grows taller than 2.5 m if pruned. Judicious pruning encourages large, full flower clusters). The flowers are borne in profusion, terminally or sub-terminally, from the branches during the summer and the rains. The shrub should be pruned to 50-60 cm during the month of February. The common varieties are 'Bride' with delicate rose flowers, *rosea* with deep pink flowers and *alba* with white flowers. There are two other cultivars - 'Petite Pinkie' having red-bronze foliage, which later turns to rich green, and has flowers of rich pink colour; and 'Petite Snow' having white flowers in snowball–like clusters. These are propagated through cuttings.

Prunus jacquemontii

Family: ***Rosaceae***

Native and distributed: Baluchistan to Chitral, Gilgit and Kaghan

Common Name: Dwarf flowering cherry

This is a small straggling shrub. It bears solitary, nearly stalk less pink flowers during May-June. It is propagated through cuttings.

Rosa sericea

Family: ***Rosaceae***

Native: China, Bhutan, Northern India (Sikkim) and Myanmar

Common Name: Thorn rose

This is a stiff erect shrub with usually hairless twigs often unarmed, with small leaflets. The flowers are solitary white or cream in colour, axillary and somewhat drooping. The flowers appears during May-August. It is propagated through cuttings. Other important temperate species are *R. moschata, R. macrophylla, R. webbiana, R. foeida* etc.

Spiraea betulifolia

Family: ***Rosaceae***

Native: Temperate Northern hemisphere or Northeast Asia, Japan

Common Name: Meadowsweet

This is a group of ornamental shrubs, which flowers to perfection at medium and high altitudes. This species is a Sun-loving dwarf (0.5 to 1 m) shrub, but in the scorching Sun of the plains the flowers wither in no time. The white flowers are borne in dense panicles during June-July. The brown-tipped anthers provide a good contrast against the white corolla. It is propagated from seeds, cuttings or layering.

Thespesia lampans

Family: ***Malvaceae***

Native: Asia

Common Name: Ban kapas, jungli bhindi

This is a shrub up to 2 m tall with large yellow bell-shaped flowers with dark crimson centers, and with usually bi-lobed leaves. the flowering season is during August-October. A weak rope can be made from the bark of this plant. The roots and fruits are employed as a remedy for gonorrhea and syphilis. It is propagated through seeds.

Vitex negundo

Family: ***Verbenaceae***

Native: South Europe

Common Name: Chaste-tree, Hemp-tree, Monk's pepper tree

This is a large deciduous shrub with whitish hairy branches and aromatic odour. Flowers are small, pale mauve in colour and fragrant. These are borne in clusters on a long terminal branch during summer and the rains. It is easily propagated through cuttings and air layering.

Weigela florida

Family: *Caprifoliaceae*

Native: Eastern Asia (Japan)

Common Name: Weigela

This deciduous shrub is grown for its showy bell-shaped flowers, and spreading and arching growth habit. Leaves are oval and tapered, to 2.5 inches long, and dark green in colour. The flowers are bell or funnel–shaped, dark pink in colour with pale pink-white interiors. They are produced in corymbs on short lateral twigs of previous year's growth in late spring to early summer. These flowers attract hummingbirds. These shrubs look very nice in a mixed shrub border. These are propagated by softwood cuttings in late spring through early summer or hardwood cuttings in mid-summer.

MAINTENANCE

Plants are like people; if you give them everything they need when they are young, they will grow up to take care of themselves very nicely.

The most crucial aspect of keeping shrubs healthy is giving them adequate water, especially when they are young.

Fertilization promotes good colour, stimulates growth, and enhances the ability of a plant to withstand adverse environmental conditions and diseases. Fertilization of the shrubs depends on soil, the kind and size of the shrub and where it is planted. If you have good soil, with lots of organic matter, you may not need any fertilizer. Chemical fertilizers are high in nutrient levels, but this can burn the shrub. Woody foliage shrubs benefit from higher proportions of nitrogen, whereas shrubs grown for flowers and berries benefit from higher percentage of phosphorous. Organic fertilizers feed the shrub and soil, but are low in nutrients. Smaller shrubs need more fertilizers than larger ones. If the shrub is planted in grass or under an established tree, it will need more fertilizer, since grass and trees take nutrients for themselves. As the plant grows, you may need to prune some branches. A general rule of thumb is not to prune more than one-third of the plant at a time. This leaves enough foliage on the plant to conduct photosynthesis, which supplies the plant with food for new growth. Some broad-leafed evergreens such as Cherry, Mountain Laurel, Aucuba, Box wood, Holly and Rhododendron can be subject to winter burn from dehydration due to water loss in the case of cold and dry winter. Care should be taken in the usage of anti-desiccant such as 'Wilt-Purf' spry, which should be applied two to three times during the winter months throughout the period of cold and thaw.

Insects, Pests and Disease Management

Most of the insects and pests, of flowering shrubs have different habits, are active at different time of the growing season, cause different types of injury and may require a specific host for their development) which attack flowering shrubs which attack flowering shrubs. Some of the important insect pests of shrubs are:

Red spider or two spotted spider mite (*Oligonychus ulmi*): This creature attacks various kinds of prunus and pyrus species plants. This creature lays red eggs near the base of young shoots during winter. They hatch out into tiny yellow red mites, which feed on the under surfaces of the leaves, generally near the midrib. They suck the sap and the foliage turns yellow in consequence. In extreme cases or later in the season, the leaves may go silvery.

Control: Spraying with white-oil emulsions, either alone or combined with rotenone or thiocyanate, is effective. Dusting with sulfur is effective in summer but in extremely hot weather it may burn the leaves of tender plants. Aerosols containing tetraethyl pyrophosphate, tetraethyl dithiopyrophosphate, or parathion are practical and effective against red spiders.

Ornamental shrubs

Pyracantha

Kalmia latifolia

Lagerstroemia indica

Aucuba

Abutilon

Lilac

Luculina

Osbeckia napalensis

Paeonia suffruticosa

Philadelphus coronarius

Potentilla

Quince

Fig. 24. Ornamental shrubs

Leaf-Mining Insects: These insects feed on the plant tissue between the upper and lower leaf surfaces and cause the formation of blotch-like or irregular serpentine mines. Insects that produce the blotch-like mines are the boxwood, burdock and arborvitae leaf miners. The columbine and holly leaf miners make serpentine mines. Some forms such as the azalea and lilac leaf miners, mine the inner tissue of the leaves for a short period, abandon the mine, roll or tie the leaves together, and feed on, or mine the opposite leaf surfaces. The boxwood leaf miner is a very small fly whose larvae or maggots feed inside the leaves. Their mines appear as blotches or blisters on the lower surface of the leaf. When they grow in huge numbers, they kill the leaves and thus disfigure the plant.

Control: Just before the adults appear in spring, spray the under side of the leaves with parathion wettable powder or chlordane in the form of an emulsion or a wettable powder. It is advisable to add a spreader-sticker to the chlordane spray to cover the shiny or waxy surface. One thorough treatment should suffice for the season.

Ants: Ants are annoying and occasionally injurious in flower gardens. Sometimes they damage plants by nesting among the roots and exposing them to drying. Some ants may also carry off newly planted seeds. Sometimes they are indirectly injurious through their habit of colonizing and protecting aphids, mealy bugs, and certain scale insects. They are often merely annoying, being attracted to plants by the presence of aphids or other sucking insects, which are excreting honeydew. Ants are also attracted by souring sap from the tree and plant wounds, and by sweet secretions of certain parts of the plants, such as the flower buds of peonies.

Control: Ants are difficult to control. Entirely satisfactory remedy is not available that will serve under all conditions and for all species. Many species are controlled by chlordane, used as a wettable powder or emulsion spray or as a dust in the manner and dosage suggested by the manufacturer. Some species can be controlled with rotenone dust.

Weevils: There are number of leaf-eating weevils such as *Otiorrhynchus sulcatus*, *Phyllobius oblongus*, *Cionus scrophulatiae*, and *Cionus pulchellus*. These weevils eat the leaves of flowering shrubs and particularly those of Rhododendrons, Buddleias and various pyrus species. The young, white, grublike larvae feed on the rootlets, and later girdle or strip the bark from the larger roots. Throughout the winter, these mostly appear as nearly full-rown or as pupae. The wingless adult females emerge in June and July. There is one generation emerging per year.

Control: Spray the parts of the plant that are above the ground with arsenate of lead, chlordane, or BHC in the form of emulsions or wettable powders late in June or early in July to kill the adults. This will prevent feeding on the bark and oviposition. Poison baits containing calcium arsenate, bran, and molasses are also useful. To control the ground-inhabiting grubs, the sprays or powders made of chlordane must be mixed with the soil.

Scales and Mealy Bugs: There are a number of scales which are found on the flowering shrubs. Sometimes the scales are conspicuously being white and may be seen on the upper and lower surfaces of the leaves. Usually the scales are brown, very much like the Mussel Scale, which attack the apple and these are found on the Cotoneasters and various species of pyrus. The hawthorns, pyracanthas and pyrus can easily be attacked by the Nut Scale which is similar to the White Peach Scale, except that the sides of the scale often protrude beyond the base.

Mealy Bugs are similar to scale insects only their bodies are protected by a waxy white secretion rather than the scales. They move about on shrubs and from plant to plant.

Control: The standard treatment is to dust the plants with sulfur either alone or mixed with pyrethrum every 10 days. Apply miscible oil or oil-emulsion sprays in the early spring while they are dormant. Two applications of a summer oil are effective, one at the time the young ones are hatching and another 10 days later. In fall, the old stems that contain the overwintering eggs should be cut off and burned to prevent reinfestation the next spring.

Aphids or Plant-lice: They infest all sorts of shrubs. Usually, they do not kill the plants but reduce plant vigour and curl or distort the leaves, harden the buds, or cause malformation of the flowers. They usually occur in colonies or clusters on the new growth, at the base of buds, or on the under sides of the leaves. Infested plants often are visited by large numbers of ants and other insects that feed upon the honeydew excreted by the aphids. They excrete a sweet, sticky, liquid that often coats the leaves or objects immediately below the aphids and gives them a varnished or sooty appearance, which results from a sooty mold that develops on the honeydew. Aphids are soft-bodied insects, whitish, greenish or blackish in colour. They have pear-shaped or nearly globular bodies and comparatively long legs. Most of the aphids serving as vectors of viruses that are harmful to plants.

Control: Spray or dust infested plants with nicotine sulfate, rotenone, or pyrethrum. Chlordane, in the form of dust or spray, is effective; so is spraying with tetraethyl pyrophosphate or lindane. To control root-infesting aphids, the following procedure has been reported to be effective:

Make a shallow trench around the base of the plants. Pour enough chlordane emulsion into it to soak the ground and reach the infested roots. Then replace the soil in the trench.

Diseases of Shrubs

There are few fungal diseases that attack different types of flowering shrubs. Fortunately, these are not very serious. It is always shrubs that have been badly looked after or are crowded or are suffering from lack of moisture at their roots that seem to be prone to fungus trouble. Some important shrub diseases are:

Mildew: The mildew is recognized by the white dust or powder, which is seen on the leaves. Sometimes it is whitish grey, but it always appears on the foliage and sometimes on the young wood as well.

Control: Spray the moment the trouble is seen with a colloidal sulphur wash. Apply the diluted liquid so as to get an even distribution all over the leaves.

Dieback or Coral Spot (*Nectria cinnabarina*)**:** When a branch has been attacked by this fungus, it dies and the death occurs right from the tip of the branch to the point of infection. It then produces tiny little pink worts or cushions on the dead wood. Unfortunately, it starts to live on the live wood and then goes on living on the subsequent dead wood. It produces its spores to reinfect other bushes.

Control: Cut out all dead wood immediately and burn it as soon as possible, or paint them over with thick white lead paint.

Shoestring Fungus (*Armillaria mellea*)**:** This is sometimes known as honey fungus. It attacks the roots of shrubs producing long strands of mycelia, which look like black leather shoestrings (as the name suggests). After a time toadstools or similar fungi may grow up from the base of the plants. This occurs at the dying stage.

Control: Never allow any dead wood to remain in a flowering shrub border. Dig up decaying stumps and burn them. By digging a narrow trench 1.5 ft deep, 3 ft away from each infected plant prevents the extension of the shoestrings.

CLIMBERS

Climbers are as essential part of a garden as are the trees or shrubs. In fact, they deserve a place even in a small garden as their growth is more vertical than horizontal. They add height, colour and texture to any garden and particularly the flowering types make great addition to any garden with a fence, trellis, wall or pergolas for them to clamber over. Climbers are a good way to cover walls and fences with flowers and foliage. But they are not all demure clingers. Some are whopping monsters as they spread quickly spreading all over. In a small garden, one must choose the right type of climber.

Climbers are plants that either have some means of supporting themselves or need external support to reach greater heights for sunlight and air. They have either tendrils, curling leafstalks or adventitious roots, which can be used for support. The climbers that do not have any inbuilt system to support themselves are actually twiners from a botanist's perspective. They just twine or coil around the support, even cover it and reach the top. In addition, some shrubs that climb are also commonly called climbers or climbing shrubs (*Jasminum humile*). They add beauty, fragrance and colour in a striking way to the gardens. They are also used to screen or decorate artificial structures such as walls, pergolas, pillars, cascades, trellis or arbours and topiary. They can be used as shade to shelter a room or patio from wind and Sun. They can also be grown in a hanging basket. They also provide food and shelter, and convenient nesting areas. Climbers from a gardener's view are of two types - annual and perennial ones.

CLASSIFICATION OF CLIMBERS

Climbers can be classified according to their growth habit and their purpose:

1. **Showy flowering climbers**: Climbers of this group generally display a show of colour with their flowers at certain times of the year. The viewer is compellingly fixed on the beautiful sight.
2. **Climbers with scented flowers**: A climber may be attractive because of its beautiful flowers and foliage or the scent of its flowers. But some are fortunate to possess both beautiful flowers and also delicious fragrance. Some of the climbing jasmines, *Trachelospermun jasminoides*, Honeysuckle, possess sweet scents. These are generally grown near the houses.
3. **Climbers with attractive foliage**: Certain climbers are mainly grown for their foliage and the flowers of such plants are insignificant or inconspicuous. Ivy and *Parthenocissus himalayan* are generally grown for their foliage only.
4. **Shade-loving climbers**: Though most of the climbers like to grow in full sunshine and produce their flowers only if they have plenty of light, there are a few climbers which can thrive under semi-shaded or shaded conditions.
5. **Deciduous Climbers**: Deciduous climbers shed their leaves during winter. Some common ones are *Campsis grandiflora*, *Clematis* spp., *Jasminum* spp.
6. **Annual Climbers**: They are seasonal plants. Sweet Pea (Lathyrus), Trailing Nasturtium, Ipomea, and Clemetis are the common annual climbers.
7. **Flowering in several flushes during the year**: Rosa spp. (Climbing Rose), *Thungbergia grandiflora*, *Vallaris heynelli* and *Trachelosprmum jasminoides*.

Climbers can be further divided into two groups according to the amount of vegetative growth they make-light climbers and heavy climbers.

Heavy Climbers: They produce luxuriant vegetative growth and grow vigorously. They cover a large area and are suitable for covering big walls or terraces. They require regular pruning to keep them in bounds. The time and method of pruning varies from species to species. Some common heavy climbers are *Jasminum* spp., *Wisteria sinensis*, *Campsis* and *Thungbergia alata*.

Light Climbers: They are such climbers that do not grow very vigorously and do not require regular or severe pruning. Some common ones are *Gloriosa superba*, *Lonicera japonica*, *Clematis* spp.

Climbers are also classified as deciduous and annual climbers.

SELECTION OF CLIMBERS

There are different locations in a garden such as sunny, partially shaded or shaded, and different structures according to which suitable types of climbers can be selected from an available wide choice. In addition, the purpose of planting the climber in any particular condition is also an important factor in deciding a suitable climber. The climbers for different conditions and locations are as follows:

For sunny conditions: *Aristolochia, Clematis, Campsis, Thungbergia alata, Jasminum dispermum, Trachelospermun, Wisteria* etc.

For partial shade: *Lonicera japonica, Trachelospermum jasminoides, Hedera helix* (Ivy).

For fragrance: Jasminum, *Trachelospermum jasminoides*, Clematis (many coloured flowers), *Lonicera japonica* (Honey Suckle), *Stephanotis floribunda*, *Wisteria sinensis*, *Passiflora cerulea* (Passion Flower-purple or red Flowers).

Climbers for pots: For growing climbers in pots, light climbers or those with bushy growth should be selected. Climbers should be planted in large pots (20-30 cm) and given suitable support to allow the plant to grow. Important climbers for pots are *Thungbergia alata*, *Gloriosa*, Ivy, *Clematis x cartmanii* 'Joe', *Hydrangea anomala petiolaris* (climbing hydrangea), *Parthenocissus*, *Stephanotis*, *Senecio,* Ivy geraniums.

Climbers for indoors: Some shade loving climbers like Ivy, Asparagus can be easily grown in pots and can be kept indoors also.

LOCATION OF PITS IN THE GARDEN

Different climbers thrive in different conditions, some preferring full Sun, but most liking at least some shade. However, they also like their roots to be in cool, shady soil, so while planting in full sunlight ensure that the root area is under planting or flagstones. The pit should always be at least one feet away from any wall to avoid the dry zone. The reason for this is that the soil near a building is usually shallow due to the foundation. Consider the height or length of climbing roses when planning to use them in the home garden. Roses climbing on a trellis or fence may be planted closer. Sunlight is important to roses and they need six to seven hours of direct unfiltered sunlight daily.

Pit and Mixture Preparation

As in the case of large trees or shrubs, climbers also need initial preparation of the soil. Most of the climbers prefer a soil having good drainage. For smaller shrubs or climbers, pits with 3 ft width and 2 ft depth should be dug up. The soil should be well mixed with 5 to 10 kg of cattle dung manure, 30 gm of sterilized bone meal, half kg of neem cake and 10 gm of some soil insecticide such as Fenvalerate or Chlorpyriphos dust to check the attack of termites. This dusting must immediately be followed by watering to settle down the soil. The soil should have good drainage. If the soil is not having good drainage, it is important to improve the soil. To do so, add bulky organic matter and large amounts of coarse sand after digging the entire soil up to a depth of 4-6 ft. It will be beneficial if a handful of bone meal is added to each climber at the time of planting. The digging up of pits or beds should start just at the beginning of the rains, so that the dug-up soil settles down with the first two or three showers.

PROPAGATION

Propagation of climbers is an important aspect. In nature plants multiply themselves by means of seeds and different vegetative parts. In a garden, a gardener has to learn the different methods of propagation by which the plants are multiplied either to replace an old or diseased stock or to meet the increasing demands of an expanding garden. Broadly, the climbers are generally propagated by two ways either sexually (seeds) or asexually (vegetative).

Sexual Propagation

Few of the ornamental climbers are propagated through seeds. Raising of climbers from seeds is a simple and economical way to propagate. One disadvantage of propagation from seeds is that the seedling may take years to produce flowers. Collect and clean ripe seeds or buy some and store them in a dry place before sowing. The seeds of some climbers have a form of dormancy that prevents them to germinate in adverse conditions. This dormancy can be broken by scarification or stratification. Sow seeds of hardy climbers in a cold frame in autumn and in open ground for the ones that do not need daily attention. The climbers which are propagated from seeds are *Aristolochia griffithii, Ipomoea versicolor, Solanum wendlandii* and *Trapaeolum majus, Pelargonium peltatum.*

Vegetative Propagation

Most of the climbers are propagated vegetatively. There are many ways to raise new plants through vegetative means such as taking cuttings, layering, division or bulbs and grafting. Taking cuttings is a simple way of propagating many climbers and is mainly useful for cultivars, hybrids and spores. Layering and division is also used to some extent. Propagation by taking cuttings can be done with different type of cuttings.

(i) Softwood cuttings

Many ornamental climbers in the home garden may be propagated by softwood cuttings. Softwood cuttings are taken from late May to early July from the current season's growth. Cutting material should be flexible but mature enough to snap when sharply bent. *Pelargonium peltatum, Trapaeolum majus* are some of the climbers that may be propagated from softwood cuttings. A proper rooting medium is needed to successfully root softwood cuttings. The rooting medium must not only retain moisture but also drain well and provide physical support. Coarse sand, perlite, and vermiculite are good rooting materials. When taking cuttings, remove the plant material with a sharp knife. Softwood cuttings should approximately be 4- 6 inches long. Pinch off the leaves on the lower half of the cutting. Make a fresh cut just below the point where one or two leaves are attached to the stem (node) and then dip the base (cut end) of the cutting in the root-promoting hormone. Insert the cutting approximately 2 inches deep into the rooting medium. Cover the cuttings to reduce water loss and then place the cuttings in bright light, but not direct sunlight. Rooting of most deciduous shrub species should occur in 6 to 8 weeks.

(ii) Greenwood cuttings - Also suitable for deciduous climbers. These can be of two types:

1. Nodal cuttings or heel cuttings; taken in late spring or early summer.
2. Semi-ripe cuttings - Suitable for evergreen and some deciduous climbers; taken from mid to late summer or early autumn. These cuttings are of two types: mallet cuttings or leaf-bud cuttings.

(iii) Hardwood cuttings - Suitable for many deciduous and some evergreen climbers; taken from late autumn to mid winter.

(iv) Root cuttings – Only suitable for some climbers; taken in mid to late winter.

Propagation by layering: Layering is a method of propagating climbers whereby a stem is encouraged to develop roots before being removed from the parent plant. Following are the types of layering:

- Air layering - Suitable for climbers with branches that are difficult to lower to the ground level; done in spring e.g. *Aristolochia griffithii, Clematis x jackmani*
- Tip layering - Suitable only for few climbers, done in spring
- Serpentine layering - Used to increase the stock of deciduous climbers; done in the dormant season

Propagation by division: Division is a simple way of increasing climbers that produce rhizomes. E.g. *Gloriosa*.

Propagating by grafting: Grafting is one of the most important ways to propagate climbers that are difficult to root or are unlikely to grow from seeds. There are three methods used for this technique:

1. Saddle grafting: Done in late winter or very early spring
2. Spliced side-veneer grafting: Can be done in summer or winter
3. Apical-wedge grafting: Done in mid winter

PLANT MATERIAL AND PLANTING SEASON

The plant material may be in the form of seeds or plants but the source of supply must be very reliable. It is heart-rending to get an indifferent bloom after tending and caring for the plant and anxiously waiting for it for months or even a year or two. It is helpful to consult the experts while making a selection. One of the convenient ways is to visit the flower shows, meet the exhibitors and the organizers, discuss with them, and make the selection. It is better to plant during early monsoon because by September, it starts to get colder at nights, so the planting operation should be carried out by the end of August. Also it should be remembered that so many of the climbers in the hills are deciduous and should be planted during the winter when they are dormant. Regular watering should be done after planting. Proper support should be provided at the time of planting. General training of the climbers in the first year is necessary to have the desired effect depending upon the kind and shape of support.

MAINTENANCE

After planting the climbers, regular watering, weeding and hoeing is required. In case of any mortality, replanting should be done. In the subsequent years, it may become necessary to prune these climbers to keep them in limit and desired shape. General pruning consists of removal of dead wood and excessive growth. Heavy climbers need drastic pruning whereas light climbers require light pruning. Deciduous climbers are usually pruned during winter months. While some are pruned in the rainy season, others are pruned just after the flowering. For example, Thungbergia is pruned after its flowering is over and Clematis is pruned lightly during the rains.

DESCRIPTION OF CLIMBERS

Fragrant Climbers

Jasminum dispermum

Family: ***Oleaceae***

Native: Himalayas

Common Name: Not known

This is a woody evergreen climber, with many fragrant pinkish-white flowers in terminal and axillary pendulous clusters. It has pinnate leaves with leathery leaflets. It flowers from April to May. It is propagated from cuttings. Other important species of this climber areJ. grandiflorum, and *J. officinale* etc.

Lonicera japonica

Family : ***Caprifoliaceae***

Origin : Japan

Common Name : Honeysuckle

This is a quick growing climber, and very popular for its sweet-scented flowers, which appear in large numbers during summer and the early rains. The tubular fragrant flowers are white in colour, tinged with purple but turn to yellow with age. It does well in the morning sunshine or in partial shades. It looks very attractive if trained along the trellis of a verandah. It is easily propagated from cuttings and by layering. Other species of this climber is *L. glabrata.*

Polygonum aubertii

Family: ***Polygonaceae***

Native: China

Common Name: Silver lace vine

This is a fast growing deciduous vine that may grow to a height of 10 to 15 ft in a single season. The foliage is bright green, but the new leaves have a reddish bronze hue. Small, fragrant flowers bloom in summer, ranging from greenish to white colour, sometimes with a pink tinge. Flowers are borne abundantly on slender panicles. The fruit is not showy. It spreads vigorously by rhizomes and is, tolerant to dry soils, almost a weed in some places. Japanese beetle may attack this plant. It has value as a quick cover, particularly where other vines cannot grow.

Rosa brunonii* Syn, *Rosa moschata* var. *nepalensis

Family: ***Rosaceae***

Native: Himalayas, Nepal

Common Name: Himalayan Musk Rose

Himalayan Musk Rose is a stout climber with small curved prickles. It has and 5-7 elliptic to oblong-lacelike leaves that are compound and finely toothed. Flowers are beautiful, fragrant and white in colour and consist of 5 petals that form a single cup. Flowers occur in clusters at the end of the branches. The fruit is purple-brown or dark red in colour, ovoid, 1 cm in diameter, smooth and shiny. It is propagated by cutting or grafting.

Schizandra chinensis

Family: ***Magnoliaceae***

Native: China, Korea and Japan

Common Name: Magnolia Vine

This is a hardy deciduous climber that grows 20 ft tall. It flowers from April to May. Flowers are fragrant, pale pink in colour, and appear singly or in clusters on long stalks. Fruits are like red berries in a grape-like cluster. The plant likes some shade with moist, well drained soil. Its berries are used in traditional Chinese medicine and valued for their ability to preserve a youthful appearance. Propagation is from cuttings of half-matured shoots in August.

Schisandra grandiflora

Family: ***Schisandraceae***

Native: China, Bhutan, N India, Nepal

Common Name: Large flowered Schisandra

This is a rare and deciduous, leathery leafed, twining climber. The flowers are delightful, beautifully scented and ivory-white in colour. The flowers are 2-3 cm or so in diameter, and are borne on trailing stalks in early summer (May-June). It is a superb climber for a shady position. It requires shelter from cold winter winds and support to climb. It is propagated from seeds and cuttings of half-ripe wood.

Stephanotis spp.

Family: *Asclepiadaceae*

Origin: Madagascar, China, Japan

Common Name: Madagascar Jasmine, Wax flower, Hawaiian Wedding Flower

The Madagascar jasmine is widely cultivated for its long-lasting waxy white flowers and intense fragrance. The scent is strongly reminiscent of the true jasmine, *Jasminum officinale,* hence the common name. It is an evergreen climber. Its trumpet-shaped blooms are in season year-round, and are a popular component of bridal bouquets. It is a vigorous climber with tough stems and dark green leathery leaves, which grow in pairs at regular intervals along the vine. It thrives best with bright but indirect sunlight. It is a vine with twining growth habit. Remember not to move the plant when in bloom, not even twist it around in the windlow as then the buds may fall. It is propagatedby cuttings or by the seeds, which are produced irregularly. Pruning back the softwood in spring encourages flowering. Also all dead or damaged wood should be removed. Important species for hills are *S. nana, S. yunnanensis, S. japonica* and *S. lutchuensis.*

Trachelospermum lucidum Syn. *T. jasminoides*

Family: ***Apocynaceae***

Native: China

Common Name: Star Jasmine

This is usually a tall woody evergreen climber, sometimes creeping over rocks, with shining elliptic pointed leaves. Flowers are strongly fragrant, white in colour, tubular, and are in loosely 3-branched clusters that appear during summer. Its main distinguishing feature is that it has both pinnate and trifoliate leaves on the same plant. It is propagated from cutting and by layering during the rainy season.

Trapaeolum majus

Family: *Tropaeolaceae*

Origin: South America

Common Name: Climbing Nasturtium

This is a fast growing annual type of climber. The bright green leaves are peltate and are 2-7 inches (5-17.5 cm) across. The plants climb by using coiling petioles. They are of very easy culture and are great as hanging basket plants, providing winter colour in the cool greenhouse. In the greenhouse, the plants bloom all winter through late spring. The 2.5

inch (6.5 cm) wide flowers come in shades of yellow, orange and red. They have a long spur and are mildly fragrant and very showy. It needs full Sun, well drained soil and cool temperature. The plant resembles the dwarf nasturtium plant in flowers and foliage. It can be easily propagated through seeds and softwood cuttings.

Vallaris heynei Syn. *V. dichotoma*

Family: ***Apocynaceae***

Origin: Himalayas, Burma and Sri Lanka

Common Name: Bread flower, Dudhi ki bel

This is an evergreen, strong-growing climber with pretty foliage. Stems are covered with a gray, spotted bark, emitting white latex when wounded. The climber bears white fragrant flowers from December to April. Bread flower vine can be trained as a compact bush and is good for container culture. It is propagated by layering.

Wisteria chinensis

Family: ***Leguminosae***

Origin: China

Common Name: Chinese Wisteria

This is best climber for cool areas. It is a woody deciduous climber. The fragrant mauve or deep lilac flowers appear in thousands during the summer months (April-May), when they are objects of great beauty. It is a heavy, woody climber, up to 30 m long in extreme cases. The creeper may be trained on walls, porches, trellis or trees. To control growth, severe pruning may be necessary. It is propagated by seeds, cuttings or layering.

Evergreen Climbers

Aristolochia griffithii

Family: ***Aristolochiaceae***

Origin: China, Himalayas

Common Name: Calico flower, Duck flower or Dutchman's pipe

This is a tall climber with large heart-shaped stalked leaves. It has very distinctive brown-purple, spotted flowers, with curved trumpet-shaped corolla with a large rounded mouth. These flowers blooms in the rainy season. It grows well in sunny conditions and is suitable for trellis. It is propagated from seeds or by layering. Cuttings taken with a heel also root in a propagating frame.

Euonymus radicans vegetus

Family: *Celastraceae*

Native: East Asia (India, China, Japan, Korea)

Common Name: Winter creeper

This is a woody evergreen vine that grows up to the height of 20 m. Similar to ivy, it climbs by means of small rootlets on the stems and also has a sterile non-flowering juvenile climbing or creeping phase. It reaches high enough into the crowns of trees to get more light, and then develops into an adult or flowering phase, which does not have climbing rootlets. The leaves are arranged in opposite pairs, elliptic to elliptic-ovate, 2-6 cm long and 1-3 cm broad, with a finely serrated margin. The flowers are inconspicuous, 5 mm in diameter, with four small greenish-yellow petals. The fruit is a four-lobed pale green pod-like berry, which splits open to reveal the fleshy-coated orange seeds, one seed in each lobe. It is used as a ground cover and a vine to climb walls and trees. This plant propagates from mature flowering stems.

Hedera nepalensis Synonyms: *H. helix* var. *himalaica*

Family: ***Araliaceae***

Native: Nepal, Bhutan, Afghanistan, India, China and Thailand

Common Name: Himalayan Ivy

This is a woody evergreen climber that grows to a height of 30 m, climbing by means of aerial roots. Leaves are simple lanceolate to ovate, entire or variously lobed, leathery, glossy and dark green in colour. Flowers are tiny, yellowish-green in colour, in stalked globular umbels 1.5-2 cm across which are arranged in domed clusters to 8 cm across. It flowers from September to October or during February-March. It is easily propagated from cuttings.

Passiflora coerulea

Family: ***Passifloraceae***

Native: Central and Western South America

Common Name: Blue Passion flower

These are tendril and evergreen climbers, which can be trained over trellises or walls. The plants need pruning after flowering, which not only keeps them in shape but often help in producing more flowers. The flowering bracts are large and green in colour, and the petals are pale pink in colour. It flowers during July-September. It is propagated by seeds or from cuttings of part-matured shoots in July or August.

Pelargonium peltatum

Family: ***Geraniaceae***

Origin: South Africa

Common Name: Ivy leaf geranium

This is one of the best container gardening geraniums, because its stems gracefully cascade over the sides of the pots. They can be easily trained up a trellis, screen or wall. When planted in a hanging basket or window box, they trail naturally. Given enough light and warmth, climbing geraniums bloom throughout the year. The colour range is wide and includes single, double and semi-double forms. It likes filtered Sun and mild temperatures. They can be grown from either cuttings or seeds. They need hard pruning in spring. Forcible branching results in a bushier and more floriferous plant.

Rosa banksiae var. *banksiae*

Family: ***Rosaceae***

Native: Central and Western China

Common Name: Lady bank's rose

This is an evergreen climber that grows vigorously over other shrubs up to the height of 6 m. Unlike most roses, it is practically thornless, though may bear some prickles up to 5 mm long, particularly on stout and strong shoots. The leaves are evergreen, 4–6 cm long, with three to five (rarely seven) leaflets 2–5 cm long with a serrated margin. The flowers are small, 1.5-2.5 cm in diameter, in double form, and pale yellow in colour. The flowers appear during early spring. It requires sunny conditions for brilliant blooming. It is easily propagated from semi-ripe cuttings or seeds.

Senecio scandens

Family: ***Compositae***

Native: East Asia, China and Japan

Common Name: English Ivy

This is a climber with zig-zag, grooved branches and arrow-shaped long-pointed leaves. It bears many lax terminal and lateral domed branched clusters of bright yellow flower-heads. It has a moderate growth rate. Flowering occurs during March-April. It should not be planted in shady areas. It is propagated from seeds, root cuttings and division in spring.

Solanum jasminoides syn. *S. album*

Family: ***Solanaceae***

Origin: South America

Common Name: Potato Vine

This is a hardy semi-evergreen climber. It is a fast growing bloomer and relatively frost-hardy. It grows vigorously up to the height of 12 ft, to form a graceful intertwined climber. Flowers are pale blue and white, in colour and are borne in many-flowered sprays from June to the first frosts. It is propagated from cuttings in July or August.

Solanum wendlandii

Family: ***Solanaceae***

Origin: Costa Rica

Common Name: Giant Potato Creeper

This is a very beautiful climber, which bears clusters of flowers in spring and early summer. The flowering lasts for 2-3 months. The pale green, deeply dented foliage matches beautifully with the clusters of mauve-blue round flowers rising above them. It flowers on new growth and hence, needs pruning after flowering. It is suitable for sunny protected locations. It can be multiplied by seeds or cuttings.

Trapaeolum canariense

Family: ***Tropaeolaceae***

Native: South and Central America

Common Name: Canary Creeper

This has a very light feathery pale green foliage and bears bright lemon yellow coloured flowers. It flowers during January- February.

Deciduous Climbers

Campsis grandiflora* Syn. *C. chinensis, Tecoma grandiflora

Family: *Bignoniaceae*

Native: East Asia, China and Japan

Common Name: Chinese trumpet vine

This is a fast growing deciduous creeper. The deep orange, red and purple flowers are funnel-shaped, opening into a trumpet-shaped mouth. The climber flowers during the summer in 6-12 flowered panicles. It is easily propagated from root suckers and cuttings. The creeper is pruned during the winter when it is deciduous.

Clematis x jackmani

Family: *Ranunculaceae*

Native: Himalayas

Common Name: Clematis

This genus consists of a large number of species, of which many are good climbers. It is a deciduous climber suitable for sunny locations and for walls. It also provides rich material for indoor decorations. The creepers produce flowers only on new growth and hence, these are pruned to give five or six buds on current season's growth in early spring. It bears violet-purple coloured flowers during July-October. Propagation is from layering or inarching on hardier species or by cuttings. Other important species of this climber are *C. langeunosa* and *C. patens*. Whitehead has listed some of the best known English varieties with different coloured flowers including Fair Rose Mond with blush-white flowers, Marcel Moser with mauve and carmine flowers, Nellie of Moser with pale mauve and rose flowers, Lady Northcliffe with lavender flowers, The President with deep violet flowers, and Belle of working with pale mauve flowers. The plants can be arranged to display their charm in a variety of ways. They are equally grown at home on arbours and on pergolas, trellis, and walls or old tree stumps. Careful training and pruning are of utmost importance to achieve good results with these plants.

Gloriosa superba

Family: *Liliaceae*

Origin: *India*

Common Name: *Glory lily*

This is a rhizomatous climber. The flower has six petals as in lilies, though the arrangement of the petals is unusual. The petals are trilled. When the flowers open, only the tip of the petals are scarlet and gradually the whole flower becomes a smashing scarlet with just a touch of golden yellow at the center from which orange anthers dart out. The plant dies at the beginning of winter. It can be multiplied by dividing the rhizomes in the early spring in March-April. It flowers during May-July. It can be grown as an annual.

Hydrangea petiolaris

Family: Hydrangeaceae

Native: Japan, Korea and Sakhalin in easternmost Siberia

Common Name: Climbing hydrangea

This is a vigorous, deciduous, woody climbing plant that grows to a height of 20 m up the trees or rock faces, climbing by means of small aerial roots on the stems. It does not require any special support; it can also be grown on old trees and stumps. The height varies from 10 to 20 ft, depending upon the pruning. The dense, deep green foliage is attractive and so are the characteristic fragrant creamy-white flowers. The flowers are not in the form of rounded heads as in the case of common hydrangea but consist of flat 'plates' held out horizontally. These are exceedingly attractive. It is grown either on walls or on trellises or fences. It is easily propagated from semi-hardwood cuttings.

Ipomoea versicolor

Family: ***Convolvulaceae***

Native: Brazil

Common Name: Mina lobata

This is a hairy climbing annual, with ovate-heart-shaped 3-lobed leaves. The flowers are large, funnel-shaped and purple in colour with pale bases and often tinged with pink. They appear from June to September. It is easily propagated from seeds.

Parthenocissus himalayan

Family: ***Vitaceae***

Native: Himalayas

Common Name: Virginia creeper

This is a very large deciduous climber that grows up to the height of 18 m. It has trifoliate leaves with 3 ovate long-pointed, sharply toothed, stalked leaflets, which are shining dark green on the upper side and pale in colour beneath. The lateral leaflets are asymmetrical. Flowers are yellow-green in colour, and are in spreading flat-topped clusters. The flowers appear during April-May. It is propagated from aerial cuttings.

INSECT PESTS OF CLIMBERS

Ornamental climbers are likely to be infested with a number of pests. Pests can be controlled with the correct identification of the problem or the pest. Pests on climbers lead to different symptoms in plants, the most common of which are bleached foliage, die back of plants, tattered flowers or the visible presence of the insects. The presence of many insects also creates webs and cotton-like material on plants. Some of the important insect pests of climbers are:

Scales

Scales are one of the most serious pests of ornamental plants. There are two main groups of scales: armored and soft. They inflict damage by sucking plant juices, and lead to stunted and poor plant growth. Scales are commonly found under leaves causing yellow spots to appear on the top. The spots get larger and if insects are not controlled in time, it can lead to premature leaf drop. Scales also like to feed on the stems and trunks of plants. Depending on the species, scales vary in size from 1/16 to 1/8 inch in diameter. Preventative measures for scales are to prune the infected plant parts and to avoid over fertilizing the plant. Chemical control methods include the use of horticultural spray oils such as 'Sunspray'to be used in summar when foliage is present.

Fig. 25. Ornamental climbers

Japanese Beetles

Japanese beetles (*Popillia japonica*) are thick, oval, half inch long insects with green bodies and coppery wings. The beetles' larvae can be found in the soil in winters and resembles the larvae of scarab beetles. The Japanese beetle is greatly destructive and infests the flowers, foliage and fruit of more than 300 different agricultural and ornamental plants from 79 different plant families. The beetles feed on the plant roots and are not only difficult but also expensive to control. The insects are usually seen during summer and the control methods include weekly use of residual or contact insecticides.

Aphids

Aphids (Aphididae) are soft, small and 0.25 mm long insects that damage plants by sucking sap and forming colonies on new shoots. Depending on the variety, aphids can be yellow, green or white in colour, and can be winged or wingless. Commonly afflicted plants include impatiens, roses and citrus hibiscus. Aphid infestation leads to stunted growth, wilting and destruction of buds, which causes poor flowering and fruit growth. In cases of severe infestation, there is distortion of leaves. Aphids are most common during spring. Aphids can be controlled with a wide range of commonly available insecticides. There is also the option to use commercial brands of chili and garlic sprays for organic gardens.

DISEASES OF CLIMBERS

Ornamental climbers are affected by number of fungal and bacterial diseases. Thus, diagnosis and management of diseases of climbers is an essential part for maintaining the beauty of a place. Important diseases of ornamental climbers based on the plant parts are:

Root rot: The initial symptoms are usually on the onset of a general decline. The foliage suddenly wilts and dries during the summer months. Dead leaves usually remain attached to the infected plant. In successive years the foliage become sparse, more chlorotic, and may curl and scorch during summer. Roots show white or cottony mats after a rain or watering.

Control: Care should be taken to prevent injury around the plants by lawnmowers, construction practices or other disturbances. Drenching of soil with benomyl or carbendazim effectively controls the disease.

Wilt diseases: A large number of climbers are susceptible to *Verticillium* and *Fusarium* species. The most common initial symptom is sudden wilting in Verticillium wilt and gradual wilt usually in upper crown in Fusarium wilt. Within a few weeks an entire side of the plant and in some cases the entire plant may wilt. Symptomatic branches containing discolouration either in diffused form or in streak is common.

Control: When a plant dies due to wilt, the entire root system should be removed and the soil should be replaced before planting any susceptible plant. Soil fumigation with methyl bromide is effective. For effective management of wilt diseases, soil drenching with Mancozeb plus Carbendazim is recommended. Soil treatment with Benomyl or Carbendazim alone is also recommended for control of many wilt diseases (Bodman *et al.,* 1996).

Black Canker: The most damaging phase for a climber is the development of cankers on limbs, trunks and branches. On larger limbs, cankers usually develop on their upper surfaces as reddish brown, sunken lesions, which may remain small and cease to extend. But in many cases, these lesions continue to advance in size and darken to smoky colour. Pimples like protuberances may be abundant over the bark of blighted twigs or along the margins of the canker.

Control: The disease can be minimized by avoiding wounds and employing sound wound treatment after pruning. Application of Bordeaux paste or Chaubattia paste is recommended for wound treatment. Application of copper fungicides, Captan and Carbendazim as foliar spray can reduce the intensity of black canker (Pirone et al 1988).

Leaf spot, blotch and anthracnose: Most of these diseases start appearing in late spring or summer. Dead areas on the leaf that are well defined from healthy tissue is the characteristic symptom of the leaf spot. Dead area on the leaf that often diffuses into the healthy tissue is the typical symptom of leaf blotch. In the anthracnose, irregular dead areas appear on leaf margin, between and across and/or along the veins. These dead areas also appear on shoots and small twigs, sometimes entire leaves are engulfed.

Control: Plant vigour can be maintained by proper fertilization and spacing. Proper spacing facilitates the sunlight to reach every corner of the plant, provides proper aeration. Collecting and burning of diseased leaves often helps in reducing the severity of leaf diseases in ensuing season. Thiophanate methyl, Chlorothalonil, mancozeb, metalaxyl, benomyl, copper hydroxide and paclobutrazol have been found effective against leaf spot, blotch and anthracnose pathogens (Tilt et al 1995, Jones et al 1996).

Powdery Mildew: During spring to late summer, foliage develops patches of grey-white coating, which resembles powder or dust on top of the leaves. These patches enlarge and often may cover the entire leaf. In some cases a white powder may be dislodged from the leaf during handling. The patches are most abundant and develop most rapidly on leaves in the lower and heavily shaded branches. The shoots and flower buds are also attacked and the infected buds fail to open (Pirone et al 1988).

Control: Pruning out infected twigs, raking up infected leaves, improving air movement around the infected hosts and increasing the amount of sunlight reaching them often reduce the severity of the disease. Dusting sulphur or wettable sulphur,which was previously used to control powdery mildew has largely been substituted by new fungicides, which are more efficacious. A spray of benomyl, carbendazim reduces the disease.

Crown gall: Crown gall disease caused by *Agrobacterium tumefaciens* attacks most of the climbers. Individual or clumps of several small swellings or knobs appear on the stem near the soil line or on roots. The galls may also appear higher on the stem of the plants. Initially, the galls are soft but later become hard and uneven. Young affected climbers die within a year or two while large established plants show gradual decline.

Control: Dipping or soaking of plants in antibiotic solutions can add protection against infection during transplanting.

SUMMARY

In recent years, ornamental shrubs and climbers have received wide attention in home garden because of their aesthetic values. These are the most important garden plants not because of the large number of cultivated species and varieties but also due to their wide range of variation in shape and size of the plants, leaves, growth habits, size, shape and colour of flowers, bark and berries. The shrubs show great variation in heights and thus plants can be chosen for growing in various places according to their suitability and utility. In addition to their utility as flowering plants shrubs are also used for hedges and screening. On the other hand, climbers decorate artificial structures like walls, arches, pergolas, pillars, cascades, topiary etc. Shrubs and climbers provide food and other life-supporting commodities in the home garden and are very important for survival of human beings and other organisms. They also protect our environment and maintain the balance in the nature. A garden lover would like to have healthy plants by following proper cultivation practices. Varied macro and microclimatic conditions make them prone to attack of large number of pathogens, which alone account for considerable damage. Severity of these pathogens can be minimized by correct identification of the problem and by modifying cultural practices, growth medium and rescheduling the pesticide applications.

A garden in a home is a place of beauty and relaxation and therefore, let the choice of plants, design, harmony and balance be as one would like them and not as per any other gardener. Their pleasant fragrance, as well as their beauty plays a role in this selection. The wild edible plants with high diversity are widely distributed in mountain forests and are valuable source of food and medicines for domestic and commercial purposes. Although there are only a few plant species fit for hills, and most of them are abundant in wild or naturalized environment, if planted properly they make good skyline, and superimpose the beauty of the surroundings.

REFERENCES

1. Ali M, Malik AR and Sharma KR 2008 Vegetative propagation of *Berberis aristata* DC. An endangered Himalayan shrub, *J Medicinal Plants Res.* **2** 374-377
2. Bodman K, Carson C, Forsberg L, Gough N, Hughes I, Parker R, Ramsay M and Whitehouse M 1996 *Ornamental Plants: Pests, Diseases and Disorders*, Queensland: Department of Primary Industries Information Centre, 158 pages
3. Grindal EW 1963 *Everyday Gardening in India* (17th edition), Bombay: D.B. Taraporevala Sons & Co. Private Ltd., 190 pages
4. Jones RK, Chase AR, Garber MP, Hudson WG, Norcini JG and Bondari K 1996 Pest management in the United States greenhouse and nursery industry II, Disease control, *HortTechnology* **6** 200-206
5. Laurie A and Ries VH 1950 *Floriculture- Fundamentals and Practices* (2nd edition), London: McGraw-Hill Book Company Inc., 257-294 pages
6. Pirone, Pascal P 1978 *Diseases and pests of ornamental plants* (5th Edition), John Wiley and Sons, 172-175 pages
7. Renfrew JM 1963 *Paleoethnobotany: The Prehistoric Food Plants of the Near East and Europe*, Columbia University Press
8. Sati SC and Joshi Savitai 2010 Antibacterial potential of leaf extracts of *Juniperus communis* L. from Kumaun Himalaya, *African Journal of Microbiology Research* **4** 1291-1294
9. Shewell-Cooper WE 1961 *The A.B.C. of flowering shrubs* (4th edition), London: The English Universities Press Ltd., 191 pages
10. Pink A 2004 Gardening for the Million, Project Gutenberg Literary Archive Foundation
11. Trivedi PP 1983 Home gardening New Delhi: ICAR, 294 pages
12. Tilt K, Goff WD and Hagan AK 1995 Effects of fungicide spray on foliage of southern magnolia, Magnolia 30 22-29.

HEDGES, EDGES, TOPIARY AND BONSAI

Desh Raj

HEDGES

Introduction

Hedge is a dense, permanent, continuous row or line of living woody or semi-woody perennial-plants grown at regular intervals and is frequently trimmed to meet the purpose of a fence, wall, screen or wind break in the garden. Hedge is an important adjunct in every garden, planted in front, sides or backyard. Hedge is primarily grown for aesthetic or utility purpose in the garden. The basic functions of hedge are security, privacy, decoration and demarcation. Thus, a good hedge should be protective, impenetrable and neatly covered with foliage from bottom to top. Although a hedge is very useful in every garden, but it should be thoughtfully planted as it competes for water, nutrients, space, air and sunshine with other ornamentals planted in the garden.

History of Hedges

History reveals that hedges were grown since the time of Romans as excavations at Farmoor (Oxon) reveal Roman hedges were made of thorn. Rackham (1986) tracked down the information that in Britain the hedges were growing for the over one thousand years. Julius Caesar encountered the practice just outside Britain in 55 BC. Part of an actual hedge which was possibly of Hawthorn plant was excavated at Bar Hill (Dunbartonshire) when the fort was built in 142 AD. Evidences of use of hedges were available during the period of Anglo-Saxon, Medieval and Elizabethan ages. From 1750 to 1850 hedges in Britain were probably planted over about two lakh miles and were equal to the hedges planted during last 500 years in Britain. It is also reported that 227 species of trees and shrubs were used for making hedges, whose age had been varying from 75 to 1100 years.

Purposes of Hedges

- Serve as a compound wall, living fence or physical barrier
- Protection against theft, animals and trespassing
- Give shelter from strong winds and gales
- Ensure privacy in the garden
- Divide different portions in the garden like drive way, parking, etc.
- Demarcate garden from the public road and buildings
- Hide unwanted sites or spots such as lavatory, compost pit, etc.
- Protection from noise, air and dust pollution
- Serve as a background for a floral display of ornamentals
- Serve as part of the garden on its own

Selection of Plants and Varieties

The selection of the plants and varieties primarily depends upon the site (microclimate), soil type and the purpose of making the hedge.

Desired Characteristics of Plant Selected for Hedge

- Attractive foliage, flowers, fruits or bracts
- Thick foliage texture
- Dense branching pattern
- Quick growth
- Drought and frost hardiness
- Thorny (essential in protective hedge)
- Quick response to pruning, trimming or clipping
- Easy to propagate either by seeds or cuttings
- Should not attract reptiles or other animals

Classification of Hedges

- Protective hedges
    - Tall (over 1 m)
    - Dwarf (50 cm to 1 m)
- Ornamental hedges
    - Based on height
        - Tall (over 1 m)

- Dwarf (50 cm to 1 m)

❖ Based on ornamental plant parts

- Foliage
- Flowers
- Fragrance of flowers
- Fruits
- Bracts

Tall Protective Hedges

***Acacia farnesiana*:** It is an evergreen plant with fine texture, dark green foliage and dark yellow flowers. It makes a good tall hedge and is suitable for growing in low hill areas in dry localities.

***Bougainvillea* species:** It is an evergreen plant. Different varieties of bougainv-illea are valued for beautiful varied colours of bracts and foliage. It is good to be grown as tall, protective and ornamental hedge in low and mid hill areas, which do not experience prolonged frost during winter.

***Caesalpinia sepiaria*:** It is a deciduous arching, tall shrub with prickly brown stem. The flowers are bell–shaped, yellow in colour and are borne in about 30 cm long racemes. It is good for growing as a tall protective hedge in mid hill areas.

***Carissa carandus*:** It is a semi-deciduous plant with glossy green foliage, thorny stem, pinkish flowers and purplish to deep violet fruits. It is good to be grown as a tall hedge in low and mid hills even on dry sloppy areas.

***Duranta repens*:** It is an evergreen, tall plant with thorny stem, drooping branches and light green foliage. The flowers are blue in colour and berries are yellow-orange in colour. It is good for growing as a tall protective hedge in low, mid and high hills.

***Elaegnus parvifolia*:** It is a deciduous, multi-stemmed vertical, thorny plant with greyish-white bark and greyish green foliage. It is good for growing as a tall protective hedge in mid hill areas.

***Mimosa himalayana*:** It is a deciduous, multi-stemmed vertical plant with ribbed, prickly reddish brown bark, fine textured prickly foliage and pinkish-white to rosy-red flowers borne in globosely heads. It is good for growing as a tall protective hedge in mid hill areas.

***Rosa moschata*:** It is a deciduous, rambling shrub with prickly stem and branches and white flowers. It is good for making a tall protective hedge in mid and high hills.

Dwarf Protective Hedges

Agave americana: It is an evergreen, succulent plant with strong needle like spine on tip of every leaf and variegated leaves. It is good for making a dwarf hedge in mid and high hills, even on dry and rocky areas.

Berberis* species: *Berberis asiatica and ***B. lyceum*** are semi-deciduous thorny shrubs with yellow flowers. They are good for making a dwarf hedge in mid and high hills.

***Craetigus crenulata*:** It is an evergreen plant with multi-stemmed spiny branches, dark green foliage and white flowers. It is good as an ornamental and protective hedge in mid and high hills.

Euphorbia bojeri: It is a succulent with spiny stem. It is good for making a hedge in low hills.

***Opuntia* species:** Different species of *Opuntia* with flat spiny stem are good for making hedge in low and mid hills, even in drought prone and rocky soils.

Pedilanthus tithymaloides: It has variegated foliage on zigzag branches. It is good for growing a dwarf hedge in low hills.

***Yucca* species:** It is an evergreen, succulent with strong needle like spine on tip of every leaf and light green leaves. It is good for growing a dwarf hedge in mid and high hills even on dry, rocky areas.

Tall Ornamental Hedges

***Buddleja asiatica*:** It is an evergreen plant with greyish green foliage with light blue fragrant flowers. It is good for growing a tall ornamental hedge, even on dry sloppy areas of mid and high hills.

***Buddleja madagascariensis*:** It is an evergreen plant with greyish green foliage and orange-yellow fragrant flowers. It is good for growing a tall ornamental hedge, even in moist, partially shaded areas in mid and high hills.

***Casuarina equisetifolia*:** It is an evergreen angiosperm tree with needle- shaped dark green foliage. It is good for growing a tall ornamental hedge in low and mid hills.

***Cestrum nocturnum*:** It is an evergreen plant with dark green foliage. The flowers are greenish-white in colour and highly fragrant during night. It is good for growing a tall ornamental hedge even in drought and frost prone areas in low and mid hills.

***Clerodendrum inerme*:** It is an evergreen plant with dark green foliage and pinkish-white, mildly fragrant flowers. It is good for growing a tall ornamental hedge even in drought prone, but also frost free areas in low and mid hills.

***Cryptomeria japonica*:** It is an evergreen, coniferous tree with fine textured dark green foliage. It is good for growing a tall hedge in high hills.

***Cupressus funebris*:** It is an evergreen, coniferous tree with fine textured green foliage in drooping branches. It is good for growing a tall hedge in mid and high hills.

***Dodonaea viscosa*:** It is an evergreen plant with dark green foliage borne on reddish brown branches. It is good for fast growing tall hedge, which can be cultivated in about five years in drought and frost prone areas in low and mid hills.

***Duranta repens*:** It is an evergreen, tall plant with light green foliage, thorny stem, drooping branches, blue flowers and yellow-orange berries. *Duranta repens variegata* (two varieties with small and large leaves) and *Duranta repens* var. Golden are very good for making ornamental hedge in low and mid hills.

***Euphorbia pulcherrima*:** It is a deciduous, multi-stemmed, vertical with green foliage and orange-red, yellow, pink bracts. It is good for growing a tall ornamental hedge in sunny, well drained soils in low and mid hills.

***Grevillea robusta*:** It is an evergreen plant with dark green leabes on the upper side and greyish green beneath the leaves. The flowers are yellow-orange in colour and mildly fragrant. It is good for growing a tall hedge cum wind break in mid and high hills.

***Hibiscus rosasinensis*:** It is an evergreen plant with dark green foliage. There are varieties with different colours such as red, maroon, yellow, cream and saffron in single, semi-double or double flower forms. It is good for growing a tall ornamental hedge in frost free areas in low and mid hills.

***Hibiscus syriacus*:** It is a deciduous, multi-stem plant with vertical growing habit. There are varieties with different colours such as white, purple and pink in single, semi-double and double forms. It is good for making ornamental hedge even in frost prone and snow bound areas in mid and high hills (wet and dry temperate conditions).

***Jasminum humile*:** It is a deciduous plant with green angular branches, dark green foliage and yellow mildly fragrant flowers. It is good for making a tall ornamental hedge in drought and frost prone areas in mid hill areas.

***Jasminum officinale*:** It is an evergreen plant with green young branches, dark green shining foliage and white, fragrant flowers. It is good for growing a tall ornamental hedge even in rocky, sloppy and drought prone areas in mid and high hills.

***Jasminum primulinum*:** It is an evergreen plant with green drooping branches, dark green foliage and yellow semi-double flowers. It is good for making an ornamental hedge in mid and high hills.

***Keria japonica*:** It is a deciduous plant with prickly branches, light green foliage and yellow semi-double and double flowers. It is good for rowing an ornamental and protective hedge in high hills.

***Lantana camara*:** It is a semi-deciduous plant with prickly branches and wrinkled hairy green foliage and multi-coloured flowers. It is good for growing an ornamental and protective hedge in low to high hills.

***Lawsonia alba*:** It is a semi-deciduous plant with dark green, small foliage and white flowers. It is good for growing a tall ornamental hedge in low hills.

***Malvaviscus arboreus*:** It is a semi-deciduous plant with dark green foliage and orange-red solitary flowers. It is good for growing a tall ornamental hedge in low and mid hills.

***Melaleuca bracteata*:** It is an evergreen plant with light yellowish green fine texture foliage. It is good for growing a tall ornamental hedge in frost free areas in low and mid hills.

***Murraya exotica*:** It is an evergreen plant with shining dark green foliage and white fragrant flowers. It is good for growing a tall ornamental hedge in low and mid hills.

***Nerium oleander*:** It is an evergreen plant with shining dark green foliage and white, pink, red and dark red flowers in single, semi-double and double form. It is good for growing a tall ornamental hedge in low and mid hills.

***Sambucus nigra*:** It is a deciduous plant with light green foliage and white fragrant flowers. It is good for growing a tall ornamental hedge in mid and high hills.

***Spartium junceum*:** It is a deciduous plant with green branches bearing dark green foliage and yellow flowers. It is good for growing a tall ornamental hedge in mid and high hills.

***Taxus baccata*:** It is an evergreen, conifer slow growing tree with dark green foliage. It is good for growing a tall ornamental hedge in partially shady and snow bound areas in mid and high hills.

***Thevetia peruviana*:** It is an evergreen plant with light green foliage and yellow flowers. It is good for growing a tall ornamental hedge in drought prone areas in low and mid hills.

***Thuja orientalis* var. *compacta*:** It is an evergreen, conifer tree with fine textured dark green foliage. It is good for growing a tall ornamental hedge and topiary in mid and high hills.

Dwarf Ornamental Hedges

***Abelia x grandiflora*:** It is a semi-deciduous plant with shining bronze-green to dark green foliage and pinkish white flowers. It is a good ornamental hedge for moist areas in mid hills.

***Aucuba japonica*:** It is an evergreen plant with yellow spots on dark green foliage like Crotons. It is a good dwarf hedge for partially shady, well drained areas in mid and high hills.

***Azalea* species:** It is an evergreen plant with dusty green to dark green shining foliage and white, pink, purple, red and yellow flowers in different varieties. It is good for growing an ornamental hedge in well drained light and acidic soils in mid and high hills.

***Berberis* species:** It is a semi-deciduous plant with thorny branches, light green to dark green foliage that turns reddish-orange during autumn and bears yellow flowers. It is good for growing a protective and ornamental hedge in mid and high hills.

***Camellia japonica*:** It is an evergreen plant with shining dark green foliage and white, pink, red and multi-coloured flowers in different varieties. It is good for growing an ornamental hedge in well drained soil in partially shady areas in mid and high hills.

***Cestrum parqui*:** It is an evergreen plant with light green foliage and yellowish-green coloured, mildly fragrant flowers. It is good for growing an ornamental hedge in mid hills.

***Cotoneaster sandakphensis*:** It is an evergreen plant with reddish brown branches, dark green shining foliage and white flowers. It is good for growing an ornamental hedge even in rocky sloppy soil in high hills.

***Craetigus crenulata*:** It is an evergreen plant with multi-stemmed spiny branches, dark green foliage and white flowers. It is good for growing an ornamental and protective hedge in mid and high hills.

***Cuphia milvillea*:** It is an evergreen plant with multi-stemmed vertical, reddish bronze to green foliage and yellow, orange-red flowers. It is good for growing a dwarf ornamental hedge in moist areas in mid hills.

***Eumonymus japonicus*:** It is an eevergreen plant with multi-stemmed vertical and shining dark green foliage.

***Eumonymus japonicus mediopictus*:** It has yellow and cream variegation in foliage. It is good for growing an ornamental hedge in well drained light soils in mid and high hills.

***Forsythia viriodissima*:** It is a deciduous plant with multi-stemmed vertical, light green foliage and yellow flowers borne on leafless branches during late winter to spring. It is a good dwarf ornamental hedge in high hills.

***Fuchsia hybrida*:** It is an evergreen plant with drooping branches, green foliage and flowers of various colours like white, pink, red, purple, violet, etc. It is good for growing a dwarf ornamental hedge in well drained soil in mid and high hills.

***Galphamia gracilis*:** It is an evergreen plant with light green foliage and yellow flowers. It is good for growing a dwarf ornamental hedge even in partially shady areas in low and mid hills.

***Gardenia jasminoides*:** It is an evergreen plant with shining light green foliage and white highly fragrant flowers. It is good for growing an ornamental hedge in mid hills.

***Hamelia patens*:** It is an evergreen plant with bronze-green foliage and orange-red, mildly fragrant flowers. It is good for growing an ornamental hedge in low hills.

***Hydrangea macrophylla*:** It is a deciduous plant with coarse textured foliage and mauve-blue, pink and white flower colour varieties. It is good for growing an ornamental hedge in mid and high hills.

***Hypericum oblongifolium*:** It is an evergreen plant with light green foliage and yellow flowers. It is good for growing a dwarf ornamental hedge in partially shady areas in mid and high hills.

***Ligustrum ovalifolium*:** It is an evergreen plant with dark green and variegated foliage. It is good for growing an ornamental hedge in mid and high hills.

***Mussaenda* species:** It is an evergreen plant with light green foliage, ornamental bracts of cream, orange, yellow and red colour. It is good for growing an ornamental hedge in frost free areas in low and mid hills.

***Philadelphus tomentosa*:** It is a deciduous plant with multi-stem vertical, bronze-red branches, dark green foliage and white fragrant flowers. It is good for growing an ornamental hedge in mid and high hills.

***Plumbago auriculata*:** It is an evergreen plant with dark green foliage and light blue flowers. It is good for growing an ornamental hedge in partially shady areas in low and mid hills.

***Spiraea corymbosa*:** It is a deciduous plant with bronze-red branches, green foliage and white fragrant flowers. It is good for growing an ornamental hedge in mid and high hills.

***Tabernaemontana coronaria*:** It is an evergreen plant with light green and variegated foliage and white flowers. It is good for growing an ornamental hedge frost free areas in low and mid hills.

***Tecoma stans*:** It is a deciduous plant with multi-stem vertical, light green foliage and yellow flowers. It is good for growing an ornamental hedge in mid and high hills.

***Vitex trifolia*:** It is a deciduous plant with greyish green vertical branches, variegated and greyish-green foliage and light blue flowers. It is good for growing an ornamental hedge in mid hills.

***Weigela florida*:** It is an evergreen plant with dark green foliage and pink-red flowers. It is good for growing an ornamental hedge in partially shady areas in high hills.

***Woodfordia fruticosa*:** It is a deciduous plant with green coarse foliage and orange red honey-scented flowers. It is good for growing an ornamental hedge in drought prone, rocky sloppy areas in mid and high hills.

The use of the above mentioned plants for hedge has been described by Arora (2007), Desh Raj (1991), Desh Raj and Sehgal (1993), Desh Raj (2000a, b, c, 2004), Randhawa and Mukhopadhay (1986), Swarup (1997) and Trivedi (1990).

Preparation of Soil

The objective of making an ideal hedge can be achieved if sunshine, water and essential nutrients are uniformly available to every plant grown in the entire length of the hedge. Thus, the selection of an ideal site is essential. Secondly, the soil ground should be well prepared. Prepare a uniform trench ideally of 30-45 cm width and depth for a hedge growing to 2 m height. If the height and width of the desired hedge is more then increase the width and depth of the trench. A deep trench ensures the development of deep root system and it reduces the competition between the hedge plants for nutrients, water and space with the neighbouring plants. Keep the trench exposed for at least a fortnight so that all the harmful micro-fauna and micro-flora gets killed with scorching Sun during May-June. Fill this trench with a mixture of well rotten FYM and sand depending on the type of soil as the whole mixture should represent one part each of soil, sand and compost. Drench the mixture with chloropyriphos (0.05%) thoroughly before planting so as to avoid the ill effects of termites and cut worms.

Time of Planting the Hedge

The ideal time for planting the hedge is during rainy season (July-September). If regular irrigation facility is available, then the planting can be done during February and March. Whenever hedge is to be raised through seeds, sow seeds in the prepared trench during February-March.

Propagation of Hedges

Plants grown for hedges are raised through seeds and stem cuttings (10-20 cm length and pencil thickness). Stem cuttings should be planted in a protected area for rooting and rooted cuttings should be planted in the prepared trench for growing the hedge.

Planting Method

The seeds or rooted cuttings are planted in double rows or in a triangular pattern in the middle of the trench. Normally, seeds are sown at a distance of 15-30 cm in two rows spaced 30 cm apart. Ideal spacing for rooted stem cuttings of tall hedge is 50-90 cm (between row to row and plant to plant), while for dwarf hedge it is 20-30 cm (between rows and plants).

Irrigation

Water requirement of hedges depends upon the prevailing weather, season and soil type. As a general recommendation, irrigation at fortnightly intervals during winter and at weekly intervals during summer is ideal. In light soil, watering is done more frequently than for heavy soils. Irrigation immediately after planting on a daily or on alternate day basis for a month ensures better plant establishment.

Manures and Fertilizers

Manuring is one of the aspects of hedge cultivation, which is over looked by majority of the gardeners as they have the wrong notion or belief that manuring of hedges is a sheer wastage of resources. But, the result of regular and appropriate manuring of hedge normally speaks for itself by producing a thick network of branches with more shining foliage and better flowering than ill-manured hedges. Hedges should be manured at least once in a year preferably before rainy season with 3-5 kg well rotten farm yard manure per metre length of the hedge. Adhikary (1999) recommended that the fertilizers, bone-meal @ 50 g, potassium sulphate @ 20 g and ammonium sulphate @ 40 g per metre length of the hedge should be applied once in a year. However, Randhawa and Mukhopadhay (1986) recommended 30 g each of bone-meal, single super-phosphate and ammonium sulphate per metre length of the hedge twice in a year.

Hoeing and Weeding

Weeds compete with the plants for nutrients, water, space and air. Thus, these should be manually removed especially during early stages of hedge development by doing the hoeing once in every month. In the later stages, hoeing and weeding three to four times in a year keep the soil porous, conserve moisture and nutrients for the development of the hedge.

Stopping or Heading Back

Every plant grown in the hedge should be cut at a uniform height of 10-15 cm from the ground level within first six months of hedge establishment. This encourages development of more lateral branches for thick network. In a tall hedge, more height is required and plants can be allowed to grow until desired height is reached before heading back.

Pruning, Trimming and Training

Pruning and trimming or clipping is the removal of unwanted growth to keep the hedge in the desired shape. It should be done regularly and the fresh growth should not be allowed to grow over 15-20 cm at any time. Pruning should be done more frequently during the rainy season. Avoid having more width for the hedge at the top than the base as poor light at the bottom can lead to more inter-nodal length and sparse growth. In sparse areas of the hedge, the branches can be bent and tied for uniform growth in the hedge.

Protection from Frost

In frost prone areas, the hedge can be covered with wet shade net or Hessian cloth or jute bags during cold night especially during December-February. Sprinkling of water on the foliage at sunset time also lowers the ill effects of frost.

Plant Protection

White ant is the major pest affecting the growth during the initial establishment days of a hedge. Drench the area thoroughly with chloropyriphos (0.05%). Keep an eye on the foliage damage by insects. Spray the same insecticide to control majority of the insects. Otherwise contact the expert for advice. For control of diseases, give precautionary spray of Dithane M.45 (0.2%) and Bavistin (0.1%) alternately once in a month.

EDGES

Edges are evergreen, low growing plants having same characteristics as hedge plants are also grown to form an edge. Edge normally grows from a few centimetres to 50 cm. To distinguish an edge from a hedge, it is low growing and does not block the view in the garden.

Purpose of Planting Edges

- To direct traffic in the garden by growing it along the paths, walkways, beds, borders etc.
- To divide different areas or features in the garden like rosary, rockery, lawn, children playing area etc.

- To mark the outline of tree basins

Common Types of Edges Planted in the Garden

- Living edge of plants
- Non-living edge of concrete, bricks, stones, tiles, blocks, etc.

Plants Suitable for Edging

Dwarf, compact and fast growing woody, semi-woody or herbaceous plants which can withstand regular pruning and trimming are suitable for making edges.

***Achillea filipendulina*:** It is a semi-woody perennial plant with fine textured dark green foliage and yellow flowers. It is good for making edge in partially shady areas in high hills.

***Alternanthera* species:** It has a herbaceous to semi-woody stem with multi-colour, bronze, red, pink and purple foliage. It is very good for making compact low edging of 20-30 cm height in frost free areas in low and mid hills.

***Aquilegia x hybrida*:** It has a herbaceous stem with fine textured foliage like ferns and has flowers in various colours such as red, blue, mauve, orange, white and pink. It is good for growing in partially shady areas in high hills.

***Arundinaria variegata*:** It has a herbaceous stem with variegated foliage. It is good for growing in partially shady areas in mid hills.

***Chrysanthemum maximum*:** It has a herbaceous to semi-woody stem with dark green foliage and white flowers. It is good for planting edge in high hills.

***Cotoneaster sandakphensis*:** It is an evergreen plant with reddish brown branches, dark green shining foliage and white flowers. It is good for growing an ornamental edge even in rocky sloppy soil in high hills.

***Cuphea hyssopifolia*:** It is a semi-deciduous plant with low growing, fine textured green foliage and pink and white flowers. It is good for growing as edge in mid and high hills.

***Dianthus chinensis*:** It is a herbaceous perennial and compact growing plant that bears pink, red and white flowers. It is good for growing in mid and high hills.

***Duranta repens* var. Golden:** It is an evergreen plant with golden yellow foliage. It is good for growing edge in frost free areas in low and mid hills.

***Duranta repens variegata*:** It is an evergreen plant with small, variegated green and cream foliage. It is good for edge in frost free areas in low and mid hills.

***Hypoesetes* species:** It is a herbaceous perennial with green foliage mottled with pink, red and white spots. It is good for edging in mid hills.

***Iresine* species:** It is a herbaceous perennial with blood red and green with red foliage. It is good for edge in frost free areas in low and mid hills.

***Myrsine africana*:** It is an evergreen plant with shining dark green foliage and red flowers. It is good for planting in partially shady areas in mid and high hills.

***Ophiopogon intermedius*:** It is an evergreen plant with dark green grassy foliage and white fragrant flowers. It is good for growing in partially shady moist areas in mid and high hills.

***Ophiopopogon intermedius variegata*:** It is an evergreen plant with dark green grassy and creamy-white variegated foliage and white fragrant flowers. It is good for growing in partially shady moist areas in mid and high hills.

***Pilea muscosa*:** It is a herbaceous perennial with aluminum shaded dark green foliage. It is good for well drained soils in mid hills.

***Rhoea discolor*:** It is a herbaceous perennial with violet-purple foliage. It is good for edging in rock gardens and raised beds in low and mid hills.

***Rumex hastatus*:** It is a semi-woody perennial with light green foliage and pink flowers. It is good for edging in drought prone, rocky, sloppy areas in mid and high hills.

***Salvia splendens*:** It is a herbaceous perennial with dark green foliage and red flowers. It is good for edging in shady areas in mid and high hills.

***Sedum* species:** It is a succulent with green to dark green leathery foliage, and red and yellow flowers. It is good for edging in rockery and raised beds in mid and high hills.

***Serrisa foetida*:** It is an evergreen, dark green with creamy-white margin foliage and white flowers. It is good for edging in low and mid hills.

***Setcreasea purpurea*:** It is a herbaceous perennial with violet-purple foliage. It is good for edging in rock gardens and raised beds in low and mid hills.

***Vinca rosea*:** It is a semi-deciduous plant with light green foliage, and pink, white and purple flowers. It is good for edging in low and mid hills.

The use of the above mentioned plants for growing edge in hills has been described by Desh Raj (1991, 2000a,c, 2004), Randhawa and Mukhopadhay (1986).

Cultivation of Edges

Cultivation of edges including preparation of soil, planting time and propagation is similar to the cultivation of hedges. However, edges are normally planted closely (10-15 cm spacing between two rows and two plants) in single, double, multiple rows or as per the triangular system. After one month of transplanting, the terminal growing shoot (2-5 cm) should be pinched to overcome apical dominance and to encourage more side branching. Regular application of manures and fertilizers, watering, pruning and trimming should be done on the same principles as for hedge to maintain the edge in good health and proper shape. For its successful cultivation, frost and plant protection measures should also be followed on the analogy of hedges.

TOPIARY

It is an art of training plants into different artistic and beautiful shapes of birds, animals, articles or statues. The word topiary is derived from the Roman word *'ars topiara'* meaning ornamental gardening. It is reported to have been initiated in first century A.D. (Randhawa and Mukhopadhay 1986). Although, an age old art, it has become a common feature only in the modern day parks and gardens. It provides the passive recreational opportunity to the visitors especially children.

Characteristics of Plants to Form Topiary

- Should be evergreen
- Should have small foliage
- Should have thick foliage density
- Should be able to withstand quick and severe pruning or trimming or clipping
- Should regenerate quickly after pruning or trimming or clipping

Plants Suitable to Form Topiary

***Bignonia unguiscati*:** It is an evergreen climber with bronze emerging foliage, light green mature leaves and yellow flowers. It is good for training in prepared structures for topiary in mid hills.

***Buxus sempervirens*:** It is an evergreen plant with dark green foliage. It is good for making topiary in high hills.

***Clerodendrum inerme*:** It is an evergreen plant with dark green foliage and pinkish-white mildly fragrant flowers. It is good for growing tall ornamental topiary even in drought prone, but frost free areas in low and mid hills.

***Cupressus sempervirens*:** It is an evergreen, coniferous tree with fine texture green foliage. It is good for growing topiary in mid, high hills.

***Duranta repens*:** It is an evergreen, tall plant with thorny stem, drooping branches with light green foliage. *Duranta repens variegata* (two varieties with small and large leaves) and *Duranta repens* var. Golden are very good for making ornamental topiary in low and mid hills.

***Ficus benjamina*:** It is an evergreen plant with dark green and creamy-white variegated foliage. It is good for growing topiary in frost free areas in low and mid hills.

***Jasminum primulinum*:** It is an evergreen plant with green drooping branches, dark green foliage and yellow semi-double flowers. It is good for making topiary in mid and high hills.

***Ligustrum ovalifolium*:** It is an evergreen plant with dark green and variegated foliage. It is good for making ornamental topiary in mid and high hills.

***Lonicera japonica*:** It is an evergreen climber with dark green foliage and yellowish-white fragrant flowers. It is good for making topiary in mid and high hills.

***Melaleuca bracteata*:** It is an evergreen plant with light yellowish green fine textured foliage. It is good for making topiary in frost free areas in low and mid hills.

***Taxus baccata*:** It is an evergreen, conifer slow growing tree with dark green foliage. It is good for making tall ornamental topiary in partially shady and snow bound areas in mid and high hills.

***Thuja orientalis* var. *compacta*:** It is an evergreen, conifer tree with fine textured dark green foliage. It is good for growing tall ornamental topiary in mid and high hills.

The use of the above mentioned plants for topiary has been described by Arora (2007), Das (2005), Desh Raj (1991, 2000c, 2004), Desh Raj and Sehgal (1993), Randhawa and Mukhopadhay (1986) and Swarup (1997).

Framework for Topiary

Developing frames for topiary requires technical and artistic hands. The frames modeled and arranged are true to the original figure of the topiary (bird, animal, article or statue) to be developed. These frames or models are made from soft iron rods or galvanized steel wires resistant to rust. Alternatively, they should be duly painted to avoid rusting later on.

Fixing Topiary Framework

The prepared topiary framework should be fixed in the garden at a suitable site (preferably sunny location) as per the requirement of the plant to be trained. The frame should be properly fixed in the ground within concrete by digging a pit (50 x 50 x 50 cm), so that it is not uprooted by blows of strong winds and gales .

Planting of Plants for Topiary

Depending upon the size of topiary, dig 2 to 8 pits (50 x 50 x 50 cm) at different points for growing the plants. Mix well rotten FYM @ 10 kg per pit and refill the pit at least a month prior to the planting. Drench 10 litres of chloropyriphos (0.05%) solution per pit to kill harmful microbes. The planting is normally done in the rainy season or during February-March.

Maintenance and Care of Topiary

Heading back of the main stem should be done at 10 to 15 cm height after the plant gets established to obtain a network of branches. To achieve the desired shape, regular training, tying and trimming of branches is required on the same principles as discussed for hedges. Mix well rotten FYM @ 10 kg per pit in addition to 50-100 g NPK mixture to the soil once in a year, preferably during December-February. Hoeing and weeding should be done 3-4 times in a year. Watering once in a week during summer and twice-thrice in a month during winter is ideal. Plant protection should be followed, as discussed for hedges.

BONSAI

Bonsai is an art, which expresses the beauty of natural tree forms in miniature. The word 'Bonsai' is comprised of two words '*Bon*' means a tray or shallow container and '*sai*' means to grow. Thus, bonsai means something growing in a shallow container or tree in a pot. *A bonsai is not a genetically dwarfed plant and is not kept small by cruelty in any way.* In fact, it is given adequate supply of water, air, light and nutrients. A properly maintained bonsai should outlive a full size tree of the same species. The techniques of Bonsai-culture are no more cruel than any other horticultural endeavour. It is also a common belief that bonsai are only a few centimetres tall. However, it is not true. Although bonsai are small in comparison to their huge life-sized specimens, most are over 15 cm to 1.5 m in height.

It is reported that the bonsai culture originated in China during 1133 A.D. However, Japanese claim to be its originators and disseminators. It reached the height of popularity in the 17th century among the Japanese. Now it is considered as a part of domestic, artistic life of the Japanese. There is hardly any house without a bonsai in Japan. All over the world, with increase in multi-storey buildings and flats, the use of bonsai for interior decoration has tremendously increased.

Bonsai Versus a Pot Plant

- A pot plant is usually valued for ornamental flowers, foliage, fruits or bracts.
- In bonsai, the beauty of the entire plant along with its harmony with the container is the matter of aesthetic appreciation.

Artistic Specialty of Bonsai

- Bonsai culture requires:
  - Painter's eye
  - Poet's imagination
  - Sculpture's hand
- Bonsai is essentially the art of choosing a plant, which has the potential of becoming a good bonsai
- It needs to be grown under complete, constant and affectionate care so that it blends with the container in which it is planted to express its natural beauty
- Thus, every bonsai, like human beings, has its own personality and individuality.

Prerequisites for Successful Bonsai Grower

- Bonsai should be watched with constant attention and affection like children, following the theory of 'Love begets love'.
- A person must spare sometime on a daily basis to look after the bonsai.

Factors Affecting the Success of a Good Bonsai

- Selection of right plant species
- Constant, complete and affectionate care
- Suitable growing media
- Adequate sunshine, water, ventilation and fertilizers
- Careful trimming, training, pruning, wiring and repotting
- Balance among the roots, trunk, branches and foliage should always be maintained

Categories of Bonsai with Respect to Size (height)

- Mamie/ Tiny - up to15 cm
- Small - 15-30 cm

- Medium - 30-60 cm
- Large - 60-150 cm

Styles of Bonsai

- **Formal upright:** Main stem is growing upright with all branches almost uniformly distributed in all directions.
- **Informal upright:** Main stem is growing upright with some branches more on one side than the other.
- **Wind swept:** Main stem and branches are tilted towards one side similar to when a natural tree appears under influence of strong wind.
- **Slanting:** Main stem is slanting and most of the branches are growing upright from the stem.
- **Semi-cascade:** The bonsai plant is grown in columnar upright container. The main stem and the branches are drooping over half of the container.
- **Full-cascade:** The bonsai plant is grown in columnar upright container kept on a stool, card board, etc. The main stem and the branches are drooping below the container.
- **Multiple trunks:** More than one trunk appears from the base and they grow upright with branches spreading in all directions.
- **Group planting:** More than one plant is growing in the same container like a tree and a shrub or a climber.
- **Raft:** The main trunk has been trained like a fallen tree on the ground and the branches appear like the stem growing upright from the fallen trunk.
- **Broom:** The main trunk is growing upright with all branches arranged in such a way that the shape appears like a broom kept upright.
- **Rock-grown:** The roots and stem of the plant are wrapped around the rock or it appears as if the plant is coming out of the rock.

Ideal Growing Medium for Bonsai

- Coarse, well drained medium, which provides basic needs such as water, oxygen and nutrition.
- Equal portion of soil, leaf mould and crushed bricks or sand is an ideal medium.
- Top layer must have sufficient humus.

- Conifer plants require more dry soil and fruit trees require soil with more humus.

Nutrition for Bonsai

- Sludge or well rotten cow dung slurry
- Take 1 kg groundnut and cotton or *neem* cake each and mix it in five litres of water (Desh Raj 2008). Allow this mixture to rot or ferment for about a month and then dilute it five times. Apply a mug of this mixture to the bonsai twice a month.
- A pinch of bone meal and single super phosphate mix is also very beneficial.

Training and Pruning of Bonsai

Training and pruning of a bonsai is necessary to give a good attractive shape to the plant.

- Growth can be reduced by regular trimming and pruning; also growth can be restricted and a proper balance between roots and shoots can be maintained.
- In one trimming not more than on-third portion of roots or shoots should be cut.

Watering of Bonsai

- A bonsai requires plenty of water and nutrients to express their maximum beauty of flowers and foliage.
- Living in a restricted space these are completely dependent on their attendant for meeting requirement of water and nutrition.
- Best time for watering is morning or evening.

Selection of Plants for Making Bonsai

- Lot of knowledge, perseverance and experience is required to select a right plant.
- Plants with smaller flowers and fruits should be selected as foliage automatically gets reduced to about one-fourth.
- Plants bearing flowers on leafless branches are very good.
- Plants selected must be able to grow in stress conditions of small growing medium and low nutrients.

Propagation Techniques

- Seeds

- Cuttings
- Layering
- Grafting
- Buying from nursery
- Collection from forests and fields

Plants Suitable for Making Bonsai in Hills

Trees

Acer oblongum, Acer pictum, Araucaria bidwillii, Araucaria columnaris, Butea monosperma, Callistemon lanceolatus, Cassia fistula, Cedrus deodara, Cryptomeria japonica, Cuppresus arizonia, Cupressus lusitanica, Erythrina suberosa, Ficus benghalensis, Ficus benjamina, Ficus religiosa, Ginkgo biloba, Grevillea robusta, Jacaranda mimosaefolia, Juniperus chinensis, Kigellia pinnata, Koelreuteria oaniculata, Melia azedarach, Michelia champaca, Persea duthiei, Persea odoratissima, Pinus roxburghii, Pinus wallichiana, Platanus orientalis, Podocarpus macrophyllus, Prunus cerasoides, Punica granatum, Pyrus pashia, Quercus leucotrichophora, Salix babylonica, Salix tetrasperma, Sapium sebiferum and *Thuja orientalis* var. *compacta.*

Shrubs

Aucuba japonica, Azalea species, *Berberis* species, *Bougainvillea* species, *Camellia* species, *Cestrum diurnum, Cestrum nocturnum, Cestrum parqui, Cocculus laurifolius, Combretum nanum, Coriaria nepalensis, Cotoneaster sandakphensis, Craetigus crenulata, Debregeasia hypoleuca, Dodonaea viscosa, Duranta repens, Elaegnus parvifolia, Erythrina blakei, Euonymus japonicus, Hiptage benghalensis, Holmskioldea sanguinea, Jasminum humile, Lagerstroemia indica, Lespedeza stenocarpa, Ligustrum ovalifolium, Malvaviscus arboreus, Mimosa himalayana, Murraya koenigii, Murraya paniculata, Nerium oleander, Osyris arborea, Viburnum cotinifolium, Spiraea corymbosa, Vitex negundo, Vitex trifolia, Weigela florida* and *Woodfordia fruticosa.*

Climbers

Hedera helix, Pyrostegia venusta, Rosa moschata, Quisqualis indica, Tecoma jasminoides and *Wisteria sinensis.*

The use of the above mentioned plants for bonsai has been mentioned by Desh Raj (1991, 2000a, b, c, 2004), Desh Raj and Sehgal (1993), Dey (2001), Randhawa and Mukhopadhay (1986), Swarup (1997) and Trivedi (1990).

Tools and Equipments Required for Making a Bonsai

- Plants
- Pots: square, round, oval, rectangular, heart-shaped, hexagonal or octagonal shape with one drainage hole at the bottom.
- Potting mixture: soil, sand and leaf mould (each part)
- Potting sticks
- Sieves
- Copper wire of 10 to 22 gauze
- Wire cutter
- Pruning knife and secateur
- Watering can and tub
- Turntable

Ideal Time for Planting Bonsai

- February-March
- July-August

Do's for Bonsai Cultivation

- Choose hardy plants suitable to the local climatic conditions.
- Choose simple container with adequate drainage.
- Position the plant correctly in the container.
- Do potting and repotting at the right time.
- Eliminate air pockets during potting and repotting.
- Trim or prune regularly to produce a fine network of shoots and roots.
- Keep the bonsai in open and airy place.
- Give balanced nutrition and use mild fertilizers or slow release fertilizers.
- Provide plenty of sunshine from all sides to the bonsai.
- Do thorough watering until water comes out of drainage hole.
- Check regularly for drainage as water logging is very harmful.
- Inspect often for insect-pests and diseases.
- Develop a design suitable for the species.
- Bonsai style must be well balanced and natural looking.

- Give your love and affection by proper and constant care to the bonsai.

Don'ts for Bonsai Cultivation

- Avoid choosing plants with large leaves, flowers and fruits.
- Do not choose a style unsuitable for the species.
- Avoid use of blunt cutting implements.
- Avoid unnatural objects for decoration.
- Do not be in a hurry to get flowers or fruits before getting desired shape.
- Do not mix chemicals and fertilizers in the same solution.
- Do not keep the bonsai on the ground as the roots through drainage hole may go deep in the soil.
- Do not tie wires very tightly as it leaves dents after removal.
- Do not retain wires for a long duration as it retains the permanent marks on the stem and branches.
- Avoid hard pruning in conifers.
- Do not give heavy dose of fertilizers at any time.
- Guard against freakish and dull appearance of bonsai and immediately look for the reasons such as poor aeration, inadequate drainage and nutrition.

REFERENCES

1. Adhikary RN 1999 Hedge and edge plants, in *Floriculture and Landscaping,* (eds) TK Bose, RG Maiti, RS Dhua and P Das, Calcutta: Naya Prokash, 663-671 pages
2. Arora JS 2007 *Introductory Ornamental Horticulture* (5th edition), Ludhiana: Kalyani Publishers
3. Das SN 2005 *Handbook of Ornamental Horticulture* (1st edition), Udaipur, Rajasthan: Agrotech Publishing Academy
4. Desh Raj 1991 Identification and Evaluation of Landscape Plants in Solan-Nauni area. *M.Sc. thesis,* Dr. Y. S. Parmar, University of Horticulture and Forestry, Solan 173 230, Himachal Pradesh
5. Desh Raj 2000a Identification of flower species for different agro-climatic zones, in *Horticulture Technology* (volume I), (eds) VK Sharma and KC Azad, New Delhi: Deep and Deep Publications Pvt. Ltd., 224-227 pages
6. Desh Raj 2000b Planting material for landscaping and their utilization in hills, in *Horticulture Technology* (volume I), (eds) VK Sharma and

KC Azad, New Delhi: Deep and Deep Publications Pvt. Ltd., 242-257 pages

7. Desh Raj 2000c Indigenous flower species,. in *Horticulture Technology* (volume I), (eds) VK Sharma and KC Azad, New Delhi: Deep and Deep Publications Pvt. Ltd., 329-331 pages
8. Desh Raj 2004 *Floriculture at Glance* (1st edition), Ludhiana: Kalyani Publishers
9. Desh Raj 2005 *Floriculture in Hills* (1st edition), Udaipur, Rajasthan: Agrotech Publishing Academy
10. Desh Raj 2008 *Floriculture at Glance* (2nd edition), Ludhiana: Kalyani Publishers
11. Desh Raj and Sehgal OP 1993 Evaluation of shrubs for landscaping in mid hills of Himachal Pradesh, *J. Ornam. Hort.* **1** 69-76
12. Dey SC 2001 *Bonsai-A Art of Miniature Plant Culture* (1st edition), Jodhpur, Rajasthan: Agrobios
13. Rackham O 1986 *History of the Countryside* (1st edition), London: J.M. Dent & Sons Ltd.
14. Randhawa GS and Mukhopadhyay A 1986 *Floriculture in India* (1st edition), New Delhi: Allied Publishers
15. Swarup V 1997 *Ornamental Horticulture* (1st edition), New Delhi: Macmillan India Ltd.
16. Trivedi PP 1990 *Beautiful Shrubs* (1st edition), New Delhi: ICAR.

HOUSE PLANTS

Neelofar, F.U. Khan, Z.A. Rather, I.T. Nazki and Z.A. Bhat

INTRODUCTION

Plants that are grown indoors in residences and offices are called house plants. Not all plants are suitable for indoor culture because of the requirement of specific ambience for growth and sustenance. The key to successful indoor plant culture is to select plants that are adaptable to the conditions in your home.

The history of growing plants indoors goes back to ancient times; chimney campanula (*C. pyramidalis*) was grown near fireplaces as early as 17th century. Historically, it seems that the earliest true house plants were succulents.

SCOPE AND IMPORTANCE OF POT PLANTS

Pot plants are grown indoors for flowers or attractive foliage. As most foliage plants are native to tropical regions, these can be grown successfully in India for pleasure and also for commerce within the country or abroad.

Culture of House Plants: Proper environment is essential for healthy growth of house plants. This environment includes physical factors such as light, temperature, humidity, moisture, growing medium and nutrition as well as biological factors such as pathogens (fungi, bacteria, and virus), pest insects and weeds.

Light: On the basis of light requirement, house plants may be categorized into:

Sun lovers: These are the plants that require 100% sunlight and cool winter. e.g. Rhodea, Saxifraga, Solanum, Bryophyllum, Cereus, Echinopsis, Crassula, Opuntia, Pelargonium, Sedum, Trichocereus.

Bright light: These are the plants that require 60-75% sunlight i.e. lightly filtered sun light from south-east or west facing window and three hours Sun a day from east or west facing windows. e.g Achimenes, Coleus, Hibiscus, Oplismenns, Rosa, Setcreasea, Thunbergia etc.

Good light: These are the plants that require 25-50% sunlight. e.g., Aechmea, Aloc, Aphelandra, Begonia, Campanula, Catharanthus, Cordyline, Cyclamen etc.

Light shade: These are the plants that require only 10% sunlight. These include Aspisdistra, Aglonema, Capsicum, Calthia cissus, Cyperus, Epiphylum, Hedera, Hydrangea, Monstera, Pilea, Syngonium and Zebrina.

Majority of plants can thrive in light more than specified though bright light category plants rarely tolerate full summer Sun. House plants, except the Sun lovers, should not be subjected to full summer Sun over mid-day as it can cause irreparable scorching.

The most expedient method of adjusting light intensity is to move the plant closer to or farther from the light source such as a window or to provide appropriately designed light from an electric source.

To reduce the light intensity, place a lace curtain between the plant and the window, use Venetian blinds to intercept and divert direct sunlight, reduce reflected light with a dark backdrop, shade the plant with another plant or move the plant back from the strong light source (for example, a south-facing window).

TEMPERATURE

In general, most of the house plants grow best at a day temperature ranging from 18-24°C and night temperature of about 10°C. Plants are categorized as per their temperature needs into following:

(1) Cool: These are the plants that thrive best under a temperature range of 7-13°C.

(2) Medium: These are the plants that thrive best under a temperature range of 13-18°C.

(3) Warm: These are the plants that thrive best under a temperature range of 18-24°C.

HUMIDITY

House plants do best at a relative humidity of 70-80%. An increase in humidity will help plants, however, especially if the dwelling is warm. One way to increase the humidity around the plants and maintain their healthy growth is to use humidifiers. Humidity levels can be elevated by other methods as well. An inexpensive way to raise the humidity is to place the pans of water on stoves or radiators to act as humidifiers. Misting is another option but prolonged misting should be avoided as this can promote growth of infective fungal pathogens. A general rule of thumb is that plants with low light requirement will need higher humidity levels.

WATERING

Water is among the most important requirements of a plant for survival but the required amount varies with the species and with the season. Some aquatic and sub-aquatic species can live in water but other majority species do not survive if there is too much water. Though sufficient water should always be poured into a pot to wet the whole of the soil and reach all the roots, no more should be given afterwards until the soil is fairly dry again. This process should be repeated. Obviously, the intervals between the waterings should be much shorter in summer than in winter. One method of testing the condition of the soil is to tap the pot sharply with the knuckles. If a clear ringing sound is returned, water is required. In general, it is better that in summer the given amount of water is liberal rather than too little; and in winter, during the resting period, it is better to give too little rather than too much. The main points as regards watering may be summed up thus:

Give water not by rule, but in accordance with the requirements of the plants. When the soil becomes dry, give enough to moisten it thoroughly. When the soil is very dry, no fertilizer should be dissolved in the irrigation water. Water, which is hard should never be given to azaleas, heaths, and rhododendrons.

NUTRITION

The applied nutrients must be of the right kind and in the right quantities. It is also important that soil, which provides the platform for growth is of good quality and is maintained in good condition. The three major nutrients are Nitrogen (N), Phosphorous (P) and Potash (K). N and K remain in the soil until they are used up by the plants, but Nitrogen (N) quickly drains away and becomes deficient. The major nutrients can be procured separately and applied in the quantities required. For small users, however, it is convenient to procure commercially available fertilizer mixtures. It is important to supply food (fertilizer) to the plants only in required quantity because mortality among pot-plants, as a result of misplaced kindness of applying overdoses, is enormous. Nutrient supply also needs to be adjusted to the cycle of vegetative growth, flowering and reproductive phases of the chosen house plants. Briefly, advice regarding nutrient supply may be summarized as follows:

A weak fertilizer may do good; a strong fertilizer may do much harm, and sometimes may be fatal. Give the fertilizer two or three times a week so long as it seems to be beneficial. Always watch the effect. Give it only to plants whose roots have filled their pots, not to those that have been freshly repotted, and never to young seedlings. Take care that the soil has been previously moistened thoroughly. Feed foliage plants during their active growing period and up to the time that growth begins to slacken, not afterwards. Treat flowering plants in the same way, except that feeding should not commence until the flower buds can be seen. Do not experiment on your plants with strong food. Finally, do not give credence to claims made in advertisements that certain patent fertilizers possess marvellous properties.

POTTING SOIL OR MEDIA

The most common ingredients used in potting soil are peat moss, composted vegetation, sand and perlite (for drainage). Some potting soil mixtures contain particles of vermicompost or vermiculite for imparting water retention. Commercially available potting soil is sterilized and adjusted for the pH. It is possible to reuse commercial potting soil, provided the remnants of plant roots, fungi, weeds and insects are removed by heat treatment.

A good potting medium like a good garden soil, must meet all basic requirements of the plant to be grown. It must provide the plant with support, good drainage, adequate air circulation, and retention of water and nutrients. 'Soilless mixes' meet these requirements admirably.

'Soilless mixes', sometimes also called peat-lite mixes, do not contain any mineral particles (i.e. sand, silt or clay). Their common ingredients include perlite, vermiculite, sphagnum moss and composted soft-wood bark. Soilless mixes are extremely lightweight, moisture and nutrient retentive, and usually sterilized. Sometimes potting soils or soilless mixes are further modified to meet specific requirements of certain plants. Some examples are African violet soil, Cactus soil and Orchid bark. Soilless media are maintained at a pH of 5.5-6.0.

Typical ingredients in a potting mix may include:

Peat moss – It is an organic product mined from decomposing plants, most often mosses, in a boggy area. The characteristics of the peat may vary slightly depending on the species of the plants that grow in that area.

Sphagnum moss – It is used for plants that require moist growing mediums and excellent aeration.

Bark products – It is ground and/or partially composted byproducts of the timber industry. This is a less expensive alternative to peat moss. Bark products have better aeration but they dry out faster than peat moss.

Perlite – It is white, volcanic rock that is crushed and heated. The heating causes it to expand. It is non-toxic, sterile and odourless. Perlite is used to improve drainage and aeration.

Styrofoam – It is an inexpensive substitute for perlite. The beads of styrofoam help to aerate potting mixes and serve as space fillers. Unlike perlite, styrofoam becomes compact over time.

Vermiculite – It is a very light, greyish and puffy substance derived from heated mica chips. It contains some potassium, magnesium and calcium that are released slowly. It is used to increase moisture and nutrient retention in mixes.

Water retention crystals – These are small translucent granules that can absorb water many times their weight. It is a non-toxic, non-biodegradable polymer with neutral pH and has a lifespan of about five years. Moisten these crystals before adding to an existing potting medium in accordance to the manufacturer's directions.

Coarse sand – These are large sand particles that can be added to improve drainage and the weight of a potting mix, particularly for succulent species (e.g. cacti). Only washed or horticultural grade sand should be used in potting mixes.

The bottom line is that packaged potting media are cheaper than the money and time spent on purchasing the plants.

PLANT CONTAINERS

Plant containers come in various materials, sizes and shapes, which are summerised below:

Plastic pots: These pots are sturdy, durable, impermeable and allow no water loss. These are easy to clean, light to handle and can be used with capillary watering mats made of felt. But these are prone to promote water logging and are not suitable for large top heavy plants.

Glazed china pots: These pots are very presentable but fragile and not compatible with felt watering mats.

Styrofoam or polystyrene pots: These pots are made of an insulating material, which keeps the roots warm and facilitates growth but are easily broken.

Clay pots: These pots are brittle, and need careful handling and storing. They are heavy and permeable resulting in water loss through the sides. Clay pots cannot be used with capillary watering mats with a wick.

POTTING

There are two methods for potting indoor plants:

(1) Planting directly in the container

(2) 'Double-potting' wherein a potted plant is placed in another more decorative container

When plants are potted directly in the container, the container should have a drainage hole and a tray to catch the excess water. In 'double-potting' only the inner pot should have drainage holes. The pots should never be placed directly in contact with the carpet, floor or furniture as moisture from the pots can damage them.

REPOTTING

Repotting becomes necessary to provide the plant with space for the acquired new vegetative growth and root mass. Usually one move a year into a larger pot is sufficient. Repotting should be done when the plants are growing actively otherwise roots will not resume growth in the new potting mixture.

How to repot: The standard procedure is to place one hand, palm downward, across the top of the pot with the plants main stem between your fingers. Turn the pot upside down and tap its rim gently on a hard surface. It should then be possible to full the pot off the root ball with the other hand. Plants with brittle stems or leaves obviously need to be turned out with great care.

Repotting large plants: Large plants must be laid on their side and the pot rim tapped with a wooden mallet or block to loosen it. If this fails, work the blade of a long sturdy knife around on the inside edge of the pot. Sometimes pots grip so hard that they may have to be broken to take out the plant.

Plants that prefer being pot bound

Some fleshly rooted plants like Clivias, bulbs like Hippeastrums, and a few shrubs flower more readily if allowed to fill their pots with roots because these plants tend to flower when stressed. If such plants are repotted, they make lush leaf growth and may not bloom. These plants with pot bound preference, however, appreciate additional nutrients especially potash, which encourages flower production.

Practical points for repotting

1. It is advisable to repot a number of plants in one operation to save on preparatory work of gathering new pots, potting mixture, drainage material etc.
2. Water the plant an hour or two before repotting and soak clay pots in water till no air bubble is seen.
3. When repotting, only a small increase in pot size is needed. For most house plants the following size increase are adequate:

Table 10. Pot size and its corresponding increase

Pot size	Increase
Up to 110.5 cm (4 ½ m)	12 mm (½ m)
12.5 to 25 cm (5 to 10 in)	2.5 cm (1 in)
Larger than 25 cm (10 in)	30.5 cm (12 in) and 38 cm (15 in)

4. With very fast growing plants, the pot size can be doubled.
5. Clay pots have one large drainage hole, which needs to be covered with a rock large enough to cover the hole and prevent soil falling out. Plastic pots have a number of small holes near the bottom edge. Plants are usually kept standing on shallow drip saucers to avoid water logging.
6. To gauge the amount of potting mixture to be filled in the new pot, place the old pot inside the new pot. Add sufficient mixture until the rim of the old pot so that when emptied into the new pot, the soil will be at the same level or a little lower. Put the root ball in the centre of the new pot and fill the measured mixture around its sides. To settle the mixture down, tap the filled pot a few times on a hard surface.
7. If the central stem of a plant is off centred, it can be repositioned when repotting by placing the root ball at an apropriate angle.

TOP DRESSING AND ROOT PRUNING

Top dressing is done for large plants in big containers to replenish the nutrient status of the plants without repotting. Remove the top inches of potting mixture till the roots are exposed and replace this layer with fresh nutrient rich potting mixture. In contrast, root pruning is done by removing slices of soil and roots from the sides and the base of the root ball with a large sharp knife after which the plant is repotted in the original container. This procedure is done as a last resort because it can cause severe shock.

POPULAR HOUSE PLANTS FOR TEMPERATE REGIONS

There are thousands of plants comprising of foliage and flowering plants, which can be grown as house plants in the temperate regions. These can be grown after providing suitable environment in the indoors. Some of these are briefly discussed in next page.

Table 11. Plants suited for pot culture in temperate regions

Plant	Availability	Condition for good growth	Ornamental attributes	Grown for
Abutilon x hybridum (D.Don) Mirbel	Rare	Full sun	Alternate, unlobed or palmately lobed leaves Common flower colours are red, pink orange, yellow and white.	Foliage and flowers
Achimenes Pers.	Moderate	Full sun	Leaves are thin and pointed or heart-shaped and are usually slightly hairy. Flowers are narrow tube-shaped but flare out into five rounded lobes with strong blue, purple pink or red shaded colour. White flowers, streaked with deeper coloured veining also found.	Flower
Agave americana, L.	Rare	Full sun	Fine grey foliage, green flowers with yellow anthers.	Foliage
Anemone x hybrida Paxt	Moderate	Full sun	Mid green, soft, oval, palmate and toothed leaves. Saucer-shaped, white flowers in clusters. Fruit is achenes and oblong.	Flower
Asparagus officinalis, L.	Moderate	Partial shade	Light green, smooth foliage. Lower leaves stalked, round, heart-shaped. Upper leaves stem-clasping, lanceolate, sharply toothed. Inflorescence raceme, cross-shaped with white flowers. Fruit is globular berry.	Foliage
Aspidistra elatior Blume	Moderate	Partial shade	Mid green, roughly pubiscent, short-stalked, spoon-shaped lanceolate leaves. Solitary, bell-shaped and cream flowers, Fruit is round berry.	Foliage
Aucuba japonica, Thunb	Rare	Full sun, Partial shade	Pale to mid green, mottled creamy white fine foliage. Olive green female flowers with creamy white anthers. Fruit is bright red berry.	Foliage
Azalea exbury hybrids	Abundant	Full sun	Small, deep green, leathery leaves and bright, funnel-shaped flowers. Pleasantly fragrant.	Foliage and Flowers

Plant	Availability	Condition for good growth	Ornamental attributes	Grown for
Begonia x *semper-florens* Link & Otto	Moderate	Full sun	Pale green, glossy and sometimes variegated leaves. Some, with white leaves, are refered to ascalla leafed bego-nias. Inflorescence is a cyme with star-shaped, pink flowe-rs. Fruit is capsule septicida.	Foliage and Flowers
Bellis perennis, L.	Abundant	Partial shade	Dark green, smooth,narrowly obovate or obovate spath-ulate leaves in a basal rosette. Solitary, pompon and pink to red flowers. Fruit is achenes and compressed.	Foliage and Flowers
Bergenia ciliata (How.) sternb	Abundant	Partial shade	Mid green pubiscent, narrowly obovate or obovate spathulate leaves in a basal rosette. Panicle, bell-shaped with mildly fragrant and pinkish white flowers. Fruit is in the form of fruit stalks and almost round.	Flower
Buxus semp-ervirens var. *suffruticosa*	Rare	Full sun	Fine foliage dark green above and lighter beneath. Flowers are pale green with yellow anthers. Fruit is oval three-pointed capsule opening in three valves, releasing shiny black seeds.	Foliage
Camellia japonica L.	Rare	Partial shade	Foliage fine dark green above and lighter beneath. Flowers white and pink	Foliage and flowers
Capsicum annum. L.	Abundant	Full sun	Ornamental cultivars have coloured foliage. Fruit of varied colour such as black and purple. Single flowers are off-white sometimes purplish	Foliage and Flowers
Chrysant-hemum leucanth-emum, L.	Abundant	Full sun, Partial shade	Mid green, smooth leaves in a rosette at the lower end, the upper ones are oblong. Solitary,pompon and white flower with achene and cylindrical fruit.	Flower
Coleus blumei	Abundant	Full sun, Partial shade and shade	Attractive foliage with red, green, pink, white, purple and yellow colour, insign-ifcant flowers	Foliage

Plant	Availability	Condition for good growth	Ornamental attributes	Grown for
Convallaria majalis, L.	Moderate	Full sun, Partial shade	Dark green, smooth, pairs of ovate lanceolate to elliptic, stalked basal leaves. Arching raceme inflorescence, with bell-shaped and white flowers. Fruit is a berry and globose.	Flower
Cuphea miniata	Abundant	Full sun	Leaves are oppositely arranged, narrow lance-shaped of varying sizes; long near the base gradually shortening towards the stem tips.	Flower
Cyclamen hederifolium	Abundant	Full sun, Partial shade	Leaves of varied shapes, dull or bright green or plain silver. Bears pink flowers with purple magenta v-shaped blotch at the base of each petal	Flower
Dendranthema x *grandiflorum* Kitam	Abundant	Full sun	Dark green, soft, alternate, oval to lanceolate leaves, deeply lobed solitary pompon, white, yellow, pink moderately fragrant flowers. Achene and cylindrical fruit	Flower
Euonymous fimbriatus, Wallich	Rare	Full sun	Light green, medium foliage which turns yellow, pink or red in autumn. Greenish white flower. Fruit is twining red.	Foliage
Euonymous fortunei, (Turcz) Hand.Mazz	Moderate	Full sun	Green leaves with either narrow white, wide yellow or broad white margins depending on the cultivar. Greenish white flowers.	Foliage
Euonymous hamiltonianus Wallich	Moderate	Full sun	Light green, medium foliage which turns yellow, pink or red in autumn. Greenish white flower. Top-shaped, deeply lobed, pink, aril orange fruit	Foliage
Euonymous Japonicus, Thunb	Abundant	Full sun	Dark green fine foliage and greenish white flowers, spherical fruit	Foliage

Plant	Availability	Condition for good growth	Ornamental attributes	Grown for
Exacum offine, Balf.f. ex Regel	Moderate	Partial shade	Dark green ovate foliage. Flowers are small bluish-lilac star-shaped with deep yellow stamens.Showy and mildly fragrant.	Foliage and Flowers
Fatsia japonica (Thunb.) Decne. & Planch.	Rare	Full sun or partial shade	Dark green fine foliage, creamy white flowers, fruit spherical black, globose, fleshy	Foliage
Fressia x *hybrida* Bailey	Rare	Full sun	Mid green, smooth, Narrowly sword-shaped leaves that develop in basal fans.Raceme is funnel-shaped and highly fragrant of mauve flowers. Fruit is Capsule & locucidal.	Flower
Fushia, L.	Moderate	Partial shade	Bushy upright habit. A few are trailing. Bear clusters of red flowers on a from the leaf axils.	Flowers
Geranium, L.	Abundant	Full sun, Partial shade	Leaves may be rounded, but more often serrated or lobed and quite attractive. Flowers are small (1") and cup-shaped; white, pink, red, purple or blue in colour.	Foliage and Flowers
Geranium rectum Trautv	Abundant	Full sun, Partial shade.	Dark green, pubescent, erect, branched; glandular leaves 6-9 partite, segments acute, incised; pedicles densely hairy; globose fruit	Foliage and Flowers
Geranium wallichianum D.Don	Abundant	Full sun, Partial shade	Dark, green, roughly pubescent. The leaves, usually five–pointed, are palmetely lobed; basal leaves are arranged in loose rosettes. Globose fruit	Foliage and Flowers
Gerbera jamesonii Bol. ex Adlam	Rare	Partial shade	Leaves, dark green above and sparsely to densely woolly beneath. Solitary, daisy like orange flowers. Fruit is Achene and Angled.	Foliage and Flowers
Hedra algeriensis, Hibb	Rare	Partial shade	Deep /pale green, mottled creamy white coriaceous leaves. Greenish yellow flowers and yellow berry.	Foliage

Plant	Availability	Condition for good growth	Ornamental attributes	Grown for
Hedra canariensis, Willd	Moderate	Partial shade	Leaves deep green in the centre merging into silvery grey and margined with white. Greenish yellow flowers with yellow-black berry.	Foliage
Hedra helix var *atropurpurea*	Moderate	Partial shade	Dark purplish green and fine foliage, greenish yellow flowers, globose black fruit.	Foliage
Hedra hibernica, L.	Rare	Partial shade	Leaves fine grey green with broad white margins. Greenish yellow flowers.	Foliage
Hedra neplensis, K. Koch	Abundant	Partial shade	Fine dark green foliage. Yellowish green flowers with globose orange fruit.	Foliage
Helleborus niger, L.	Moderate	Full sun, Partial shade	Dark green, fleshy, over wintering basal leaves Clusters of saucer-shaped white flowers. Fruits are follicles and flattened.	Flowers
Hippeastrum puniceum Urban	Rare	Full sun, Partial shade	Mid green, smooth, semi-ere ct, strap-shaped basal leaves. Funnel-shaped, pink flowers. Fruit is globuse capsule.	Flower
Hosta plantaginea Aschers	Moderate	Full sun, Partial shade	Light green, glossy, ovate to heart-shaped, slightly wavy leaves. Racemes are bell–shaped, strongly scented white flowers. Elongated fruit	Foliage and Flowers
Hyacinthus orientalis, L.	Abundant	Full sun, Partial shade	Dark green, fleshy, leaves. Racemes are bell-shaped. Highly fragrant violet flowers. Fruit is three angled capsule.	Flower
Hydrangea arborescens, L.	Moderate	Full sun, Partial shade	Foliage dark green above, downy beneath and of fine texture. Flowers are pink, rose purple, blue or mauve. Fruit is many seeded and dehiscent from the top.	Foliage and Flowers
Hypoestes sanguinolenta	Moderate	Partial shade	Colourful splashes of foliage with bright pink spots on dark green leaves. Hybrids with bright red color foliage	Foliage

Plant	Availability	Condition for good growth	Ornamental attributes	Grown for
			or white spotting also available. Flowers inconspicuous.	
Juniperus communis, L.	Moderate	Full sun, Partial shade	Deep green to blue green sharpely pointed fine leaves, with single, glaucous white band. Flowers inconspicuous Female cones have three fleshy, interlinked scales. Green at beginning and turn black at maturity.	Foliage
Juniperus horizantalis, Moench.	Moderate	Full sun, Partial shade	Grey green, needle like and scale like adult leaves. Flowers inconspicuous, ovoid dark blue fruit.	Foliage
Juniperus indica, Betrol.	Moderate	Full sun, Partial shade	Fine leaves, bluish above furrowed and green beneath/ on lower branches.	Foliage
Juniperus recurva, Buch-Ham ex D.Don	Moderate	Full sun, Partial shade	Fine, grey green, narrowly wedge-shaped leaves.	Foliage
Kalanchoe blossfeldiana V. Poellnitz	Rare	Partial shade	Dark green, glossy,oval-oblong ovate, softly toothed leaves. Corymb like panicles. Tubular-shaped, orange flowers. Fruit are follicles and flattened.	Foliage and Flowers
Laurus nobilis, L.	Rare	Full sun, Partial shade	Dark green foliage, greenish yellow flowers. Fruits drupaceous green; black when ripe	Foliage
Ligustrum ovalifolium Hasak.	Moderate	Full sun	Dark green fine foliage. Small fragrant yellowish white flowers borne in panicles, spherical shiny black fruit	Foliage and flowers
Lilium candidum L.	Moderate	Partial shade	Dark green, glossy, broad, inversely lanceolate basal leaves. Raceme is trumpet-shaped and flowers are white. Fruit is a capsule.	Flower
Lilium longiflorum Thunb.	Moderate	Partial shade	Mid green, smooth, sessile, atternate, or nearly opposite or whorled leaves Racemes are trumpet-shaped and flowers are white. Fruit is Capsule.	Flower

Plant	Availability	Condition for good growth	Ornamental attributes	Grown for
Lonicera nitida, Wils	Moderate	Full sun, Partial shade	Dark green foliage coriaseous above lighter beneath, bronze, creamy white fruit; connate glossy blue purple berries.	Foliage
Mahonia aquifolium (Pursh.) Nutt.	Rare	Full sun, Partial shade	Dark green leaves which turn to luminescent bronze orange to dull purple brown to wine red in autumn/ winter. Flowers are yellow. Fruit is black purple, blue primrose, with reddish juice.	Foliage
Mahonia japonica, (Thunb.)	Rare	Full sun, Partial shade	Dark green medium foliage changing to wine red in autumn/winter	Foliage
Myrtus communis, L.	Moderate	Full sun	Dark green leaves which are reddish when young. Flowers are white and fruits are small, spherical and blue black.	Foliage
Pelargoniam x hortorurn	Abundant	Full sun, Partial shade	Mid green, somewhat rounded in shape and having bronze or maroon markings. Flowers are white, pink, red, salmon, orange, violet and all shades in-between, bicolors and double flowering forms are also available.	Foliage and Flowers
Petunia x hydrida	Abundant	Full sun, Partial shade	Foliage ranges from mid to dark green. Flowers are available in both single and double forms and in colours ranging from purple, scarlet, salomon to white.	Foliage and Flowers
Phormium tenax J.R.& G.Forst	Rare	Full sun, Partial shade	Foliage coarse greenish, margins and midrib, bordered with red or orange line	Foliage
Primula denticulata Smith	Abundant	Full sun, Partial shade	Mid green, Roughly pubescent, petiole. The lower surface covered with yellowish powder. Umbel is bell-shaped. Yellow flower and cylindrical fruit	Foliage and Flowers
Primula rosea Royle	Abundant	Full sun, Partial shade	Dark green, soft, elliptic, spathulate leaves with the base narrowed into a winged petiole. Umbels are cross–	Foliage and Flowers

Plant	Availability	Condition for good growth	Ornamental attributes	Grown for
			shaped, flowers are pink and fruit is globular capsule.	
Primula vulgaris Huds	Abundant	Full sun, Partial shade	Dark green, roughly pubescent toothed to scalloped, deeply veined leaves. Umbels are rosette of pale yellow flowers. Fruit is globular capsule.	Foliage and Flowers
Prunus lanrocerasus, L.	Rare	shade	Foliage, dark green above, pale green beneath. Flowers are white.	Foliage
Punica granatum, L.	Rare	Full sun	Bright green, fine foliage turning to grey orange with coppery or red veins in autumn. Scarlet or orange red flowers. Pome is large yellow red with numerous seeds.	Foliage, Flowers and fruit
Rosa banksiae, Ait	Rare	Full sun	Pale green fine foliage and white flowers.	Flowers
Rosa brunonii, Lindl.	Abundant	Full sun	Mid green fine foliage. White flowers.	Flowers
Rosa x damascena, Mill.	Moderate	Full sun	Dull, grey green fine foliage. Pink or red flowers. Obvoid red bristle clad fruit.	Flowers
Rosa foetida, *Herrm.*	Rare	Full sun	Bright green fine foliage. Yellow flowers. Spherical red hips.	Flowers
Rosa multiflora, Thunb.	Rare	Full sun	Dark green fine foliage. White flowers.	Flowers
Rosa rugosa, Thunb.	Rare	Full sun	Dark green wrinkled foliage. Wine crimson flowers. Tomato-shaped red or orange hips.	Flowers
Rosa webbiana Royle	Abundant	Full sun	Dark green fine foliage .Pink flowers. Red ovoid to flask-shaped fruit.	Flowers
Ruscus aculeatus, L.	Rare	Partial shade	Dark green leaves.Flowers are small, white with a dark violet or greenish centre. Fruit is a red or yellow berry.	Foliage
Ruta graveolens, L.	Moderate	Partial shade	Leaves are bipinnate or tripinnate of feathery appearance and green to blue green in colour. Flowers are yellow and borne in cymes. Fruit is a lobed capsule	Foliage and Flowers

Plant	Availability	Condition for good growth	Ornamental attributes	Grown for
Saintpaulia ionantha	Moderate	Full sun	Thick, fuzzy leaves and vibrantly blue–violet flowers. Blooms throughout the year under right conditions.	Foliage and Flowers
Santolina chamaecypar issus	Abundant	Full sun	Leaves often covered with silvery hairs. Flowers are yellow. Fruit is angled.	Foliage and Flowers
Schizanthus Pinnatus		Full sun	Oblong to lanceolate deep green leaves with wavy mar-gins. Flowers are small, white and star-shaped. Fruit is small, orang-red to red in colour.	Foliage and Flowers
Sedum spectabile Boreau	Abundant	Full sun	Grey green, fleshy and, toothed leaves. Cymes is star-shaped. Pink flowers Fruit is follicles and erect.	Foliage and Flowers
Thuja orientalis L.	Abundant	Full sun	Scale like mid green or yellow green fine leaves turning bronze in autumn. Male catkins are yellow and female catkins are purplish. Fruits are ovaoid cones.	Foliage
Trades-cantia sillamontana	Moderate	Full sun	Grey green, silky leaves. Cymes are saucer-shaped and and flowers are of magenta colour. Fruit is a capsule.	Foliage
Tradesc-antia virginiana, L.	Rare	SS/FS	Mid green, fleshy. Leaves are alternate and hairless Cymes are saucer-shaped and flowers are purplish blue. Fruit is angled capsule.	Foliage
Trapaeolum majus	Full sun Partial shade	?	Bright green leaves. Orange or yellow, mildly fragrant flowers that trail gracefully over the rim of the pot	Foliage and Flowers
Vinica rosea	Moderate	?	Oval or oblong, glossy green hairless leaves with pale midrib and a short petiole, arranged in opposite pairs. Flowers are white to dark pink with darker red at the center, fruit is a pair of fothiles.	Foliage and Flowers

Plant	Availability	Condition for good growth	Ornamental attributes	Grown for
Zantedeschia aethiopica Spreng	Abundant	Full sun	Dark green, glossy, semi-erect, arrow-shaped leaves. Inflorescence is funnel-shaped and flowers are white.	Flower
Zantedeschia elliottiana Engler	Rare	Full sun	Mid green, heart-shaped basal leaves with translucent white spots. Spathes are funnel-shaped and flowers are golden yellow.	Flower

SUGGESTED READING

1. Anthony Huxley 1983 *The macmillan world guide to house plants,* Turnohout, Belgium: Henri Proost et Cie.
2. Hugh C. Davidson 1922 *Culture of pot plants in rooms, greenhouses and frames* London: Crossby Lockwood and Son. Stationers Hall Court, Ludgate Hill
3. Swarup Vishnu 1996 *Indoor Gardening* New Delhi Indian Council of Agricultural Research Krishi Anusandhan Bhavan
4. Wikipedia,the free encyclopedia.www.wikipedia.org.

ANNUAL FLOWERS

M. K. Singh

INTRODUCTION

Annuals are pivotal to the exuberance of colours in gardens and are a major motivation for a gardener to stay interested in the creative hobby of formal or home gardening. The variety and choice among the annuals is enormous and in recent years the widened repertoire has seen a tremendous upsurge because of innovative application of science. Many new varieties of annuals, from Petunias and Antirrhinums to Poinsettias and Hibiscuses, are so transformed in their physical dimensions, growth habit and colour patterns of their flowers and foliage that they are even difficult to relate to their corresponding older counterparts.

In this chapter, an attempt is made to present an overview of the choices available among the annuals, the basics of their planting and aftercare rather than exhaustive coverage of technical details. It is recommended that the gardeners should visit a good nursery in their neighborhood, see the varieties available, and ascertain their suitability and adaptability to conditions and environmental parameters under which the targeted plants are to be grown.

CLASSIFICATION

Annuals are generally classified into three broad groups according to their growing season.

Summer Season Annuals

The summer season annuals grow at higher temperatures. The seeds are sown from the last week of February to the first week of March and seedlings are transplanted in beds or pots at the end of March or in the beginning of April. The common summer season annuals are Kochia, Zinnia, Gaillardia, Tithonia, Marigold, Coreopsis, Sunflower etc.

Rainy Season Annuals

These annuals are grown during the season when there is frequent precipitation and high humidity in the environment. The seeds of rainy season annuals are sown during May-June and transplanted during June-July. Amranthus, Celosia, Torenia, Gomphrena, Balsam and all summer annuals are grown in rainy season.

Winter Season Annuals

Winter season annuals are able to tolerate relatively low temperature and are the preferred choice of gardeners for lending colour to their enterprise in the more favourable environment of winter season. Majority of the annuals are cultivated in winter months in the plains. Most of these annuals have been introduced by the British in India. The nursery of winter season annuals is raised in September and transplanted into beds or pots in October in plains and low hills; in temperate hills, these are sown during February-March and transplanted during March-April. Winter season annuals provide a beautiful display of colours in the garden and this group includes a large number of seasonal flowers. The most important winter season annuals are Pansy, Phlox, Ice plant, Candytuft, Gomphrena, Sweet William, Sweet Pea, Annual Carnation, Annual Chrysanthemum, Calendula, China Aster, Dimorphotheca, Sweet Sultan, Marigold, Dahlia, Stock, Helichrysum, Salvia, Petunia, Calendula, Antirrhinum, Hollyhock, Gypsophila, Limonium, Lupin, Poppy, Sweet Alyssum, Viola, Amranthus etc.

Annuals are a versatile group that can cater to diverse requirements and situations such as growing in beds, pots, hanging baskets and for cut flowers or loose flowers. Annuals suitable for specific use are:

1. **For Bedding:** Pansy, Phlox, Ice plant, Candytuft, Balsam, Zinnia Gomphrena Sweet William, Carnation, Sweet Sultan, Gaillardia, Marigold, Dahlia, Portulaca, Stock, Cosmos, Helichrysum, Salvia, Petunia, Calendula and Antirrhinum
2. **For Loose Flowers:** Seasonal Chrysanthemum, Marigold, Aster, Sunflower, Gaillardia and Zinnia
3. **For Hanging Baskets:** Nasturtium, Phlox, Portulaca, Daisy, Pansy, Petunia, Verbena, Sweet Alyssum, Dwarf Marigold, Zinnia and Ageratum (dwarf)
4. **For growing in Pots:** Pansy, Phlox, Ice plant, Candytuft, Balsam, Zinnia Gomphrena Sweet William, Carnation, Gaillardia, Marigold, Dahlia, Portulaca, Stock, Salvia, Petunia, Calendula and Antirrhinum
5. **For obtaining Dry Flowers:** Helichrysum, Statice, Molucella, Carnation and Acroclinum

6. **For growing in shade:** Salvia, Cineraria, Ageratum, Alyssum, Phlox, Verbena, Clarkia, Iberis and Impatiens
7. **For Cut Flower:** Sweet William, Carnation, Sweet Sultan, Helichrysum, Antirrhinum, Aster, Larkspur and Lupin
8. **For decorative Screening:** Sweet Pea, Holly hock and Helichrysum
9. **For Fragnance:** Alyssum, Carnation, Phlox, Heliotropium and Stock
10. **As Climbers:** Sweet Pea, Nasturtium and Morning Glory
11. **As Green Filler:** Mollucella, Gypsophilla and Kochia
12. **For Garlands:** Marigold, Gillardia, Chrysanthemum and Zinnia

CULTURAL PRACTICES

Propagation

Plants can be procured from market as seedlings or small plants or can be raised from seeds. Seeds of some annuals can be sown directly in the garden at the selected location. These include Morning Glory, Sweet Pea, Lupin and Nasturtium. Many other annuals do best if seedings are raised and then transplanted. Seedlings should be produced indoors, 4-6 weeks before planting in the garden.

Raising Nursery

In general, seeds of annuals are small. These are sown at high density in raised beds or pots or shallow seed pans depending on the quantity of sown seeds. The nursery beds should be of 3-4 m length, 60-80 cm width and 15-20 cm height above the ground level. The growing media should consist of one part each of soil, sand and sieved dry leaf mould or vermicompost that is moistened before sowing the seeds. The soil should be drenched with 0.2% Captan solution to protect against infection from soil borne diseases in the nursery.

Before sowing, the seeds should be treated with a fungicide such as Bavistin or Captan to protect the seedlings from fungal diseases. Seeds should be sown in lines, 3-4 cm apart at a depth of 0.4-0.6 cm. Small seeds such as those of Petunia may be mixed with a small quantity of fine sand to get uniform spread. After sowing, the seeds should be covered with sieved mixture of vermicompost, sand and soil. Water should be sprinkled carefully immediately after sowing. Thereafter, watering should be done twice a day keeping in mind that over-watering is harmful for seedlings of annuals. The beds should be covered with light cloth or newspaper to create darkness, which helps better germination. After germination, covering should be removed and the seedling should be exposed to sunlight in order to prevent the seedings becoming lanky. It takes about one month to grow the seedlings sufficiently strong and ready for transplanting.

Soil and Bed Preparation

Soil for annuals should not be acidic. It must be fertile, and well drained with good water holding capacity. If the soil is acidic, addition of lime is a must to make it neutral in nature. Sandy loam soil of 6.0-7.0 pH is ideal for successful cultivation of annuals. Addition of organic matter and sand in clay and stiff soil can improve the soil by making it porous. In case of sandy soil also, a lot of FYM should be added to improve its texture. The bed should be dug upto 30 cm depth but the digging should not be done in very wet soil. The soil working should be repeated 2-3 times to obtain the best results. Beds should be of 1.0 m width and of convenient length, and soil should be well pulverized. Before planting, mix decomposed organic manure @ 5.0-8.0 kg per square metre to the soil thoroughly.

The growth and flowering is optimal under full Sun. Salvia, Cineraria, Ageratum, Alyssum, Phlox, Verbena, Clarkia, Iberis and Impatiens can also be grown in partialshade conditions.

Transplanting

A month old seedling of an annual flowering plant is sufficiently strong and ready for transplanting. The transplanting should be done either on a cloudy day or in the afternoon. Generally, seedlings should be planted in beds at a distance of 15 x 15 cm, 15 x 10 cm, 30 x 30 cm, 45 x 30 cm or 45 x 45 cm. The beds should be irrigated soon after transplanting.

Plant density, as manipulated by spacing, has a strong impact on flowering behaviour. Kumar and Kaur (1997) observed highest plant height (55.33 cm) in *Statice cv. Forever Purple* at a spacing of 60 x 30 cm. Jhon et al (1991) recorded that spacing of 30 x 60 cm gave highest plant height (127.67cm), number of branches per plant and number of flowers per plant in *Zinnia elegans cv Giant Flowered Crimson*.

Irrigation

Since annuals are sensitive to salt, it is important that good quality water is used for obtaining better growth of plants. Rain water is invariably ideal. Salt content of the irrigation water should be less than 10 micro mol per litre. Most annuals need regular watering because they do not develop deep root systems. Containers and baskets require most frequent watering as they dry out quickly. The water requirements vary widely from area to area. Careful checking is important, even during the rainy season. Top layer of soil i.e. 20-25 cm of soil should be kept moist all the time because the development of roots is most vigorous in this region. If sufficient water is not provided to the plant, the flower production declines.

Fertilization

Well-prepared soil and organic mulch make the nutrients available to plants, but annuals grow so rapidly that supplementary fertilizers are needed to maintain vigour. Rotten FYM @ 5-8 kg per square metre should be incorporated into the soil at the time of soil bed preparation. In addition to FYM, it is advisable to apply 10-15 g per square metre nitrogen and 8-10 g per square metre each of phosphorus and potash for getting good vegetative growth and flowering. Whole of phosphorus and potash should be applied at the time of land preparation. Nitrogen should be applied in two split doses. The first dose should be applied 20-25 days after planting and the second dose 40-45 days after transplanting. Nutrient availability and supply has a strong correlation with quantity and quality of the bloom. For example, Handreck (1986) observed that stock needed 25-27 mg per litre sulphur through fertigation. Application of 40 kg N along with 200 kg P_2O_5/ha produced the highest yield of flowers of China aster (Yadav 1983). Aboul (1994) reported early flowering and more number of flowers when seeds of *Z. elegans* were sown in soil inoculated with VAM fungus *Glomus etunicatum*. Smith and Treaster (1991) observed good growth and flowering in Gomphrena with the application of municipal sludge compost at a depth of 1.5 inches and 2 inches deep mulch at the time of planting.

Pinching and Disbudding

In many annuals, because of apical dominance the plants grow tall and lanky with very less branches. Such plants are susceptible to lodging and the flower production also suffers. Pinching should be done to break apical dominance and to encourage axillary branching and thereby increasing flower production. Pinching slightly delays flowering but improves flower production. Pinching, 20-25 days after transplanting, has been found to be optimal. In case large flowers are desired for exhibition purposes, axillary buds should be removed when they are at 'Pea Size' stage.

Rajasekaran et al (1983) recorded that the Gomphrena plant produced maximum number of flowers when pinched 15 days after transplanting.

Weeding

Keep annuals free from weeds. A combination of hand weeding and mulch is effective. Weed regularly to prevent weeds from setting seeds and establishing themselves. If weeds are not controlled in time, growth, flowering and productivity are drastically reduced. At least 3-4 manual weedings are required for the full cropping period. Pre-emergence weedicides, which prevent weed seeds from germinating, have limited use in flower gardens. Phetpradap and Hampton (1991) observed that pre-sowing application of trifluralin at 2 kg per hectare significantly reduced weed population but did not affect the emergence of China Aster seedlings.

Mulching

Mulching is beneficial for the annual flowering plants in many ways. It retains the soil moisture, reduces soil erosion, controls the weed population and reduces the leaching of nutrients. Organic mulch can made with weed-free materials such as rice straw or other vegetative garden waste. Mulching is also very useful for protecting winter annuals from frost and snow.

Staking

Many tall annual flowering plants must be staked or provided with other support system, especially in windy and exposed areas. Wind, rain or the weight of foliage and blossoms can bend or break the stems of the plant and ruin the display. Dahlias, Hollyhocks and others of the kind can grow to reach 6-7 ft in height and because of the weight of heavy stalks and bloom, they may need support or staking for which material like split bamboo sticks may be used.

INSECTS

Aphids

Aphids are usually found on young leaves and flower buds. They suck the sap of foliage and bud, and cause retarded growth of the plant with poor quality flowers. Two sprays of Endosulfan or Rogor @ 1.0-1.5 ml per litre of water at 15-20 days interval can control aphids.

Caterpillars

The caterpillars are particularly active in summer. They damage the foliage and flower buds. They can be controlled by spray of Endosulfan or Malathian @ 1.0-1.5 ml per litre of water.

Thrips

The thrips suck sap from the undersurface of the leaves and also from the flower buds. Thrips thrive in dry hot summer season. Spraying of Rogar or Monocrotophos @ 0.1% controls the thrips.

Mites

Mites infect the annual flowering plants and give them a dusty appearance. The plants show webbed appearance. These can be controlled by spraying Kelthane (Dicofol) @ 0.1%.

Leaf hoppers

The hoppers suck sap from the leaves and stem. The infected plants show cupped or rolled leaves and give wilted appearance. Spraying of Rogor (0.2%) controls the hoppers.

DISEASES

The annual flowering plants are affected by a number of fungal diseases, which damage the plants and their flowers. The most destructive diseases are botrytis, root rot and foot rot.

Botrytis

In a Botrytis infected plant, brown colour spots are developed and are visible on the flower petals during the period of high humidity. It can be controlled by spraying Mancozeb @ 2.0 g per litre of water.

Damping off and root-rot

Root rot is caused by the fungus *Pythium*, which prefers moist conditions. Infected plants lose the root system and suffer heavy loss in flower production. It generally attacks the plants grown in heavy and compact soil. Root rot can be controlled by sterilization of soil before planting, better air circulation and decreasing the moisture content of the growing media. The disease can be controlled by prophylactic drenching of nursery beds with Captaf (0.1%) and Bavistin (0.1%). Seed treatment with Captaf (0.1%) also reduces the incidence of the disease.

Dressing of seeds with Captan preparation to control damping off and foot-rot diseases, results in higher percentage of seedling emergence and survival (Niebisch 1985).

Leaf spot and blight

Various pathogens cause leaf spots and blights in annual flowers. The symptoms appear as small brownish spots on the leaves. These leaf spot diseases can be controlled by spraying Dithane M-45 @ 0.2%.

Inflorescence blight

Infected inflorescence gives a burnt appearance in severe cases. This disease can be controlled by spraying Dithane M-45 @ 0.2%.

Powdery mildew

Initialy, whitish tiny spots appear on the leaves. The entire plant shows white powdery appearance in severe cases. The disease can be controlled by spraying Karathane (0.1%) or Claxin (0.03%).

SEED COLLECTION

For most families such as Leguminosae, Liliaceae, Cruciferae and Papavaraceae, seed or pod collection should be started when the ovary wall starts to change from green to yellow colour. With other families such as Compositae, Labiatae and Umbelliferae, each flower head should be harvested individually when the seed is turning brown in colour. At the time of picking, flowers in which seed setting has not taken place should be avoided. The dried flowers should be threshed and winnowed to obtain the seeds. The seeds should be stored in muslin cloth bags or paper bags in a well ventilated place.

REFERENCES

1. Aboul-Nasar A 1994 Effect of vesicular-arbucular mycorrhiza on *Tagetes erecta* and *Zinnia elegans*, *Alexandria J. Agril. Res.* **39** 453-470
2. Handreck KA 1986 Critical concentrations of sulfur in liquid feeds for plants in containers, *Scientia Horticulture* **30** 1-17
3. Jhon AQ, Paul TM and Tanbi MI 1991 Effect of nitrogen and plant spacing on growth and flower production of *Zinnia elegans*, *Adv. Plant Sci.* **4** 1-7
4. Kumar R and Kaur K 1997 *In vitro* propagation of Gerbera. *J. Ornamental Horticulture* **5** 20-25
5. Niebisch RM 1985 Machrichtenblant fur den Pflanzenschutz inder DDR **39** 190-191
6. Phetpradap L and Hampton JG 1991, 300-303. Proc. Forty-fourth New Zealand Weed and Pest Control confer
7. Rajasekaran LP, Shanmugavelu KG and Nambisan KMP 1983 *South Indian Horticulture* **36** 248-249
8. Smith EM and Treaster SA 1991 OARDC Special Circular-Ohio Agri. Res. Development Centre, No. 137
9. Yadav LP 1983 Ph D Thesis, Bidhan Chandra Krishi Vishwavidyalaya, West Bengal, India

ROSE

M. K. Singh

INTRODUCTION

Rose is perhaps the most important ornamental bedding plant in a garden and is universally acclaimed as the 'Queen of flowers'. It derives popularity because of its adaptability to various soil and climatic conditions, long blooming period and beautiful flowers with good shelf life, which lends themselves to numerous uses ranging from religious offerings to intricate decorative arrangements. Rose petals are used as medicament in Ayurveda and for extraction of highly prized essential oil used in the perfumery industry. Rose is the number one cut flower in the international market.Roses are extensively cultivated for cut flowers in Netherlands, Columbia, Japan, Italy, Germany, France, Spain, United States of America (USA), Morocco, Korea, Israel, Kenya, Zimbabwe, India and Malaysia. In flower markets, Hybrid Tea (H.T.), Floribunda and the Spray types are mainly in demand for production of cut roses.

ORIGIN AND HISTORY

The genus *Rosa* (*Rosa hybrid*) is endemic to temperate region including sub-temperate and subtropical areas of northern hemisphere i.e. North America, Europe, Asia and Middle East. In all other areas, where roses are found growing, it is from their introduction by professional growers and botanists. Broad classes of roses are Old European roses, China roses, Hybrid Perpetuals, Tea roses and Hybrid Tea roses. The genus *Rosa* includes 200 species and more than 18,000 cultivars.

From the approximately 200 species of wild roses, 11 species that grow wild in India are *Rosa clinophylla* Thory, *R. brunonii* Lindl. (Himalayan musk rose), *R. foetida* J. Herrm., *R. eglanteria* Linn. (syn. *R. rubiginosa* L.), *R. moschata* J. Herrm. (Musk rose), *R. sericea* Lindl., *Webbiana royle* (Ladakh rose), *R. gigantean* Collet ex Crep (syn. *R. odorata* (Andr) Sweet Briar, *R. leschenaultiana* Wight and Arn (syn. *R. sempervirens* L.), *R. longicuspis* Bertol. and *R. macrophylla*.

Many ornamental roses are also grown for attractive or scented foliage (*Rosa glauca* and *R. rubiginosa*), ornamental thorns (*R.sericea*) or for their showy fruits (*R. moyesii*). Ornamental roses have been cultivated for millennia, with the earliest known cultivation dating back to 500 BC in the Mediterranean countries, Persia and China. In India, roses are being grown since ancient times, mainly for beautification of houses and temples.

CLASSIFICATION

Roses are mainly grouped into eight classes: Tea, Hybrid Perpetual, Hybrid Tea, Floribunda, Polyantha, Miniature, Climber and Rambler. Their major characteristics are given below:

Teas

This profuse blooming rose is known as 'Tea' for the similarity of its fragrance to tea. It is not hardy and requires protection of polyhouse or glasshouse during the winter season. These roses prefer well drained light loamy soil. Very little or no pruning should be done for the best performance of Tea roses.

Hybrid Perpetual (H.P.)

It is tall and hardy plant and a cross between European and Asian ancestors. Hybrid Perpetuals bear moderately large flowers, which sometimes possess rich fragrance. They prefer good loamy or sandy loamy soil and can be grown successfully in places of high rainfall and humidity. Moderate to heavy pruning should be done to H. P. roses depending on the plant growth.

Hybrid Tea (H.T.)

This type owes its origin to crosses between Teas and Hybrid Perpetual. Large shapely bloom is usually produced in bunches of three or more to the stem. Common practice is to remove the side buds so that all the food materials from the stem are directed into one flower of excellent quality. H. T. roses prefer well drained loamy soil with rich organic matter and full sunshine. Medium to hard pruning should be done depending upon the variety and growth of the plant.

Floribunda

These roses have resulted from crosses between H.T. and Polyantha. Flowers are usually double and long-lasting that open almost at the same time. Floribunda roses produce flower clusters of 3-15 flowers. Plants need fairly light prunning to create a massed bedding effect.

Polyantha

Bushes of Polyantha are dwarf statured, bear clusters of small blooms that are free flowering and of perpetual bearing habit. The hardy type bushes need very little pruning. They also prefer good loamy or sandy loamy soil.

Miniature

Miniature roses grow 13-38 cm high. Recent Miniature roses have small leaves and exquisite little flowers that are borne on small bushes. Most Miniature roses are useful to be grown as edging rose beds. Miniatures are also suitable for growing in pots. These roses have evolved mainly from two species of China roses.

Climbers

Climbing roses help to dress up any garden. They have upright climbing habit and require support. They have heavy spring bloom followed by scattered blooms throughout the season. Climbing roses are a diverse group of many different heritages, which makes this a wonderfully useful collection of roses. Many of climbers are bud sports of the Hybrid Teas and Floribundas.

Rambler

Ramblers are ideal for covering unsightly buildings, high walls, pergolas and for scrambling up trees. Pruning requirements for these are different from other climbers as flowers of ramblers are produced on the previous year's growth. So, these need to be pruned earlier and straight after the main flush of flower.

VARIETIES

Successful rose gardening depends upon the proper choice of the site, and selection of classes and varieties of rose. In different classes of roses, many varieties are available which vary in plant height, growth habit, and shape, colour and size of flowers. The basic idea is to beautify the garden for enjoyment and this needs aesthetic sense. Indian varieties of roses enjoy a special position in the garden. The names of a few varieties of garden roses are given below, which the gardener will find of interest.

Hybrid Teas

Yellow

Poornima, Vasant, Ganga, Chitwan, Manu Mukherji, Santi, Pusa Sonia, Golden Splendour, Golden Giant, Landora, Lemon Spice, Summer

Sunshine, Basanti, Dr. Budhen, Dr. Homi Bhabha, Double Delight, Helmut Schmidt and Pampa

White

White Nun, Hans, Mahasweta, Meghdoot, Tushar, Sweet Innocence, Divine Light, Jawahar, Ivory Tower, Frau Karl, White Lightnin, Baijubawara, Bapuji, Meghdoot, Honor, Lemon Sherbet and Ruksaar

Pink

Apsra, Arati, Kulu Belle, Dr. B. P. Pal, Indian Princess, Lalima, Pat Rani, Nurjehan, Rajkumari, Pusa Christina, Mrinalini, Haseena, Surbhi, Sukanya, Uttam, Anurag, Radhakrishna, Lucky Lady, Royal Highness, Abhisarika, Akashsundari, Arati, Avra, Chamak, Mridula, Nazar-e-Nazar, Eiffel Tower, Heidi Jayne, Jadis, May Lyon, Montreal, Picture, Peach Beauty, Prima Donna, Rajni, Supriya, Taj Mahal and Tapti

Orange

Shyama Prasad, Nayika, Deepak Rag, Bhairavee, Modern Art and Super Star Supreme

Orange Scarlet

Ramakrishna Dev, Aruna, Raja Surendra Singh of Nalagarh, Cherry Brandy, Hawai, Queen Fabiola, Elida and Madelon

Red

Anupama, Bhim, Dark Boy, Chitrarekha, Agnihotri, First Rose Convention, Kalima, Lal Bahadur, Rat Ki Rani, Raktagandha, Sugandha, Chitralekha, Cara Mia, Happiness, Red Lion, Rose Planet, Ambica, Angar, Argha, Bidhan Babu, Cauvery, Night-n-Day, Night Time, Oklahoma and Royal William

Lavender Mauve

Blue Delight, Madhushala and Ajanta

Apricot

Delhi Apricot, Raja Ram Mohan Roy, Sun God, Balwant, Golconda, Just Joey, Whisky, Ambassador, Banaras Dawn and Maid of Honour

Bicolour

City of Panjim, Kurwat, Srinivasa, Padmavati, Surkhab, Bajazzo, Oriana, Archana, Love and Rose Gaujard

Floribundass

Yellow

Golden Roys, Mahadev, Agni Veena, Arthue Bell, Charisma and Sun Flare

White

Navneet, Shabnam, Chandrama, Akito, Laminuette, Park Place, Picasso, Pink Parfait, White Junior Miss and White Queen Elizabeth

Pink

Arunima, Amrapali, Belle of Punjab, Chamba Princess,, Kumari, Prema, Ragini, Delhi Pink Powderpuff, Kavita, Nav Sadabahar, Shrinagar, Azeez, Bridal Pink, Inner Wheel, Jiminy Cricket, June Bride, Lorena, Lutin, Queen Elizabeth, Red Gold, Rose Parade and Zorina

Orange

Deepika, Delhi Maid, Mohini, Sandhya Bela, Suryodaya and City of Leeds

Red

Nutkhut, Pandit Nehru, Jantar Mantar, First Offering, Panchu, Annabell, Cordula, Imperator and Zizi

Lavender Mauve

Neelambari, Sushma, Africa Star and Shocking Blue

Colour Blend

Kamini, Madhura, Priya, Kadambari, Banjaran, Chingari and Chitchor

Polyanthas

Bharani (white), Chattilon, Pink Showers, Priti, Rishi Bankim, Swati, Tarapunja (pink), Vatertag (orange), Anjani (red) and Nartaki (yellow)

Miniatures

CRI CRI, Dazzler (pink), Over the Rainbow (bicolour), Rise-N-Shine, Delhi Starlet (yellow), Strawberry Swirl (blends of red, pink and white), Chandrika and Pushkala (white)

Climbers

Akash Pradip (red), Climbing Dr. Homi Bhabha, Climbing Guitare, Climbing Rina Herholdt, Delhi Pink Pearl, Deoghar Clg No. 1, Deoghar Clg No. 4 (pink), Lalkila (red), Landour, Miss Harp Clg, Kanyakumari (pink), Delhi White pearl (white) and Deoghar Clg No. 5 (yellow)

Ramblers

Fountain, Jeanee Lajoie, Joseph's Coat (yellow), Orange Cascade (red), Over the Rainbow (red), Pink Cascade (pink), Red Cascade (red) and Snow Carpet (white).

CLIMATE

Roses can be grown successfully under various climatic conditions. Quality flowers are obtained in areas with bright sunny days, cool nights, moderate humidity and absence of strong winds. In general, it prefers day temperature between 25-28°C and night temperature between 16-18°C. Long days of 12 hours and humidity of 60-70% favours growth and quality flower production. During flower bud initiation, long days play an important role but in the later stage of bud development it is influenced by light intensities. By providing additional lighting and temperature, early flowering can be achieved. Bhattacharjee (1991) observed that quality rose blooms are obtained only from December-April under Delhi conditions. In hilly regions best quality rose bloom is obtained from April-November.

SELECTION OF SITE

Selection of site is very important for successful growing of roses. The selected site for planting the rose should get full day sunlight. In summer season, it prefers partial shade during the hot afternoons. In the garden, rose plantation area should not be too close to tall trees. Rose plants can not survive in water logging conditions even for a short period. The location of rose plantation should be free from dust.

SOIL AND PREPARATION

Rose plants grow well in medium loamy (Bose and Mukherjee 1977), rich in organic matter, well drained soil of at least 40-50 cm depth. The sub-soil must have the capacity to hold and retain adequate moisture. Water logging adversely affects growth therefore, clayey soils which are heavy and poor in aeration are not preferred for rose plants. In case of clay soil, it is advised to mix sand and well decomposed FYM to make it porous. The best soil for roses should have a pH of 6.0-6.5 with electrical conductivity (EC) less than 1.0. Now-a-days, artificial media such as cocopeat, perlite and mixture of cocopeat and perlite are preferred because they are inert, sterile and have no cation exchange capacity. So, the applied nutrients are available to the rose plants directly. Roses fall into the sensitive category of plants with respect to salinity and sodicity tolerance (Bhattacharjee 1995). Budded plants have less tolerance to adverse conditions of soil and climate than the plants raised from cuttings.

Roses are perennial plants and can remain in the same bed or pit for many years. The initial preparation of the soil should preferably be started during summer season. In the garden, the design of beds should be simple and informal. Rectangular beds with 30 cm wide path between two beds offer certain advantage for cultural operations. The length and width of a bed may be 6.0 x 1.2 m. In the rectangular bed, planting two or three rows is a good practice. Dig the pits, 30 cm in diameter and depth, at least one month before planting and expose them to sunlight. Few days before planting, fill the pits with one part of well decomposed FYM and one part each of fresh soil and coarse sand. In termite prone areas, antitermite chemicals such as chlorpyriphos should be applied to the soil before planting. Pits should be irrigated to settle the mixture.

PROPAGATION

Rose is propagated by seeds, budding, grafting, layering, cuttings, and through micro-propagation for disease-free plants. Asexual propagation allows plants to grow true to type. It is commercially propagated by budding.

Seed Propagation

The rose plant is propagated by seeds for multiplication of some root-stocks, for maintenance of some species of roses or for raising new hybrid varieties. Some rose varieties and wild species bear fruits known as 'hips' which contain seeds. Fully ripe fruits are harvested and seeds are extracted after drying the hip. Seed germination is often difficult because the seeds are covered with a hard seed coat. Hence, the seeds need to be subjected to scarification and stratification to a cold temperature between 1.0-4.4°C for 10-15 weeks before sowing them in pots to raise seedlings. Venkataraman et al (1990) observed maximum percentage of seed germination, where the extracted seed were stratified, scarified and treated with GA_3 at 100 ppm. At the time of transplanting, the seedlings should be of 10-12 cm height. After transplanting, it should be lightly irrigated.

Vegetative Propagation

i. Stem Cuttings

Rose root-stocks are generally propagated from matured stem cuttings. This is the cheapest method of propagation of roses. This method is particularly adopted for multiplying Miniature, Polyantha, Climber and Rambler roses. It is also useful in multiplication of *Rosa multiflora, R. bourboniana, R. moschata, R. canina and R. indica* var. Odrata as root-stocks. Usually, one year old shoots from healthy plants are used as cuttings. Rooting of cuttings varies with the type of cutting, season when cuttings are taken, growing media, moisture conditions, species, and growth regulator treatment. The cutting may have one, two or more buds. Bhujbal and Kale (1973) recorded maximum percentage of rooting in *R. multiflora,* when these cuttings were treated with 1000 ppm each of Indole Butyric Acid (IBA) and Indole Acetic Acid (IAA). The technique of raising plants from cuttings of matured shoots, in hilly region is as follows:

During winter months (September-November), one year old shoot from a healthy plant is cut from the base. From this, several pieces measuring about 20-22 cm long each are cleanly cut from just below the node with a sharp secateur and the lower leaves are removed. The cuttings are generally planted 5-6 cm deep, keeping a distance of 8-10 cm between cuttings and 12-15 cm between rows, in sand or cocopeat or mixture of cocopeat and perlite (1:1 ratio) under partial-shady conditions. Sand is excellent media for rooting (Balakrishnamuethy and Roa 1988). Watering should be done by a watering can. After 12-15 weeks, sufficient roots are formed on the cuttings and they can be transplanted in the field or into polythene bags.

ii. Budding

In this method of propagation, detached bud from a selected rose variety is inserted into selected root-stock. T-budding is the most widely accepted method for propagating ornamental rose plants. The cambium layer is the growing layer, which is found between the wood and the bark of the shoot. Success of budding mainly depends on joining the cambiums of the root-stock and that of the bud. The performance of the budded variety depends much on the root-stock in terms of vigour and production of quantity and quality of flowers. It is necessary to choose the most suitable species of root-stocks. *Rosa indica* var. *odorata* in Northern India and *R. multiflora* in Southern India have been found highly promising and are extensively used as rose root-stocks. The best time for budding in North India is during February-March, in Eastern India during October-November and February-March. In the hills, budding may be done in spring season (Randhawa and Mukhopadhyay 1986).

iii. Stenting

Stenting is a method for quick propagation of roses. A piece of stem of the cultivar with one dormant bud is grafted on a root-stock. In stenting, formation of bud or graft union and formation of adventitious roots occur simultaneously, resulting in a complete plant in 2-3 months.

Tissue Culture

In vitro, propagation of ornamental roses is important for mass multiplication of planting materials in a short time. Micropropagation of *Rosa hybrid* was successfully done in cvs Andhra Red (Vijaya and Satyanarayn 1990), Queen Elizabeth (Rout et al 1989) and Super Star (Bhat 1992) from explants of lateral bud. This method is also suitable for mass propagation of root-stock.

PLANTING

Rose plants revel in low temperature and the root growth is very active in the winter season. Its plants can be planted at any time of the year. However, irrespective of the locations in hills, October is the most suitable time for planting roses. In temperate climate, it can be planted in open field during October-November and February-March (Malik and Dadlani 1984). Rose plants are best planted in rows. Rose beds present a harmony of colour and height of plants. A spacing of 50-60 cm between plants and rows within beds may be required for Hybrid Teas, Polyanthas, Floribandas and Teas. In case of vigorous growing bushes of rose plants, the spacing may be 75-100 cm. To get maximum flowering shoots, roses should preferably be planted 20 x 30 cm apart. Roses should be planted on thoroughly prepared pits in beds. Before initiating plantation, it should be ensured that the soil in the pits is enough moist. A hole, measuring 20-30 cm in diameter and 30 cm in depth, will be sufficient in most of the cases. The depth of planting will vary with the type of planting materials. The budded plants should be planted in such a way that the budding or matrix point should remain 5-7 cm above the soil surface.

Irrigation

Roses are very fond of water and require one to two inches of it every week. During hot, dry season, even more may be needed depending on the size of the bushes. In spite of their need for abundance of water, roses hate standing in water and hence, provision for proper drainage is essential. Irrigation should be done during the morning time. Avoid evening watering as it could foster powdery mildew, a very common disease among roses. During the rainy season, watering will be required only during the dry periods between the rain events.

Pruning

Pruning is one of the most important operations in rose cultivation, which influences the yield and the quality of flowers. All dry, diseased, weak and criss-cross shoots should be removed during pruning to open up the center of the bush. Any suckers that arise from the root-stocks should also be removed. The best time of pruning is during the period when the rose plant is at dormant or near to dormant stage. It will depend on the climatic conditions of the rose growing region. Usually, the Hybrid Tea and Floribunda roses take 45-50 days to come to flowering after the pruning. In plains, pruning is recommended in October after the rainy season. In hilly areas, where the winter is very cold, the pruning operation should be done from last week of February to mid-March.

Nutrition

Like all plants, roses require air, water, nutrients and sunshine for growth and development. The yield and quality of flowers can be markedly increased through supply of proper quantity of fertilizers and manures to plants. Roses require nitrogen, phosphorus, and potassium are required in large quantities. For roses, the entire quantity of phosphorus and potash should be applied at the time of land preparation. Nitrogen should be applied in split doses. Well rotten FYM @ 5-6 kg per square metre should be incorporated well in the soil at the time of soil preparation.

When plants are nitrogen deficient, the young leaves become small in size, develop pale green colour with yellow and red tints and the buds develop poorly and become light in colour. Nijjar and Rehalia (1977) observed that in North India, a high nitrogen dose (50g per plant per year) was more effective to increase stem length, flower yield and quality of flowers of cultivar Super Star.

Phosphorus is important to encourage root growth and to stimulate flowering. Phosphorus can be applied in the form of super phosphate and phosphoric acid. Its deficiency symptoms in roses are blue green or dark green leaves, weak stems and slow bud development due to poor roots. With application of 60 g nitrogen in four split doses in combination with 20 g P_2O_5 per square metre per year in cv Montezuma produced better quality flowers (Yadav et al 1985).

Potassium is responsible for hardiness of the flower stem, better quality of bloom, and increased resistance to diseases. Due to potassium deficiency, lower leaves show browning, stem becomes weak and colour of flower gets poorer. Uma and Gowda (1987) reported that application of 16 g per plant per year of N and K_2O each, produced maximum shoot length in cv Super Star.

An investigation on the effect of nitrogen in combination with phosphorus and potassium on growth, flower yield and quality of four indigenous cut flower rose varieties (namely Pusa Gaurav, Dr. Bharat Ram, Arjun and Nehru Centenary) at Palampur (H.P.) revealed maximum diameter of flowering stem and flower bud, width of flower, number of petals per flower. Flowers with stem length of 60 cm and above were recorded with application of 9 g N + 8 g P_2O_5 + 8 g K_2O per square metre per week (Singh and Kumar 2010).

Foliar spray of micronutrients is more effective in roses. Biswas (1984) recommended that 3 g each of magnesium sulphate and potassium sulphate along with 0.75 g borax dissolved in 1 l of water and sprayed on rose, can correct the deficiency of magnesium, potassium and boron.

Weeding and Hoeing

Weeding if neglected, may cause a setback in flower yield and quality. Manual weeding is the most prevalent practice. Hoeing of rose beds at fortnightly intervals is helpful in removing weeds and loosening soil for better aeration and improving soil moisture retention capacity. Shallow hoeing is preferred as deep hoeing destroys the fibrous roots.

Mulching

Mulching reduces weed infestation and promotes economy of irrigation water. Various materials such as dry plant leaves, black polythene film, and paddy straw can be used for mulching. About 6-8 inches mulch of dried leaves of Jamun tree in rose beds in the month of April does not allow any weeds for about two months i.e. in May and June, and reduces their reappearance in following months (Singh et al 2003).

Pinching

Removal of a part of terminal growing region of shoot is known as pinching and is an important operation in rose cultivation. In the garden, after every flush, pinching is done for the succeeding return crop and for adjusting flowering for a specific timing. Pinching reduces the plant height and increases new axillary branching. Pinching should be carried out above three leaflets in weak shoots and five leaflets in healthy shoots.

Disbudding and Disuckering

Disbudding of axillary buds in Hybrid Tea roses is beneficial to increase the size of its flower. Zieslin et al (1976) observed that removal of lateral bud from young plants did not increase the number of flowers in cv Baccara but increased flower production in cvs Sonia and Belinda by 50 and 75% respectively. In Floribunda rose, removal of the central (apical) bud improves flower quality and gives the cluster a balanced appearance. The suckers that appear from root-stocks should be removed periodically from the point of their origin.

INSECTS AND PESTS

Red spider mite (*Tetranychus urticae*)

Mites are very common on rose. They first appear on lower surface of leaves, and can subsequently cover the entire leaf, stem, and after sometime the entire plant with their web. The infected plants give dusty and webbed appearance. These can be controlled by spraying Kelthane (Dicofol) @ 0.1% or Omite @ 0.03%.

Caterpillar

Caterpillars are mostly a problem of rose buds and young leaves. In the initial stages, they eat away the foliage. The eggs are laid on the buds and larvae eat into the bud and damage it completely. These can be controlled by spraying Nuvan @ 0.15 to 0.2%.

Aphids (*Macrosiphum rose* and *Aphis gossypii*)

Aphids occur in more than 4000 aphis species. Many are vectors for a number of viruses. Aphids are sucking insects that draw chlorophyll from leaves and buds. Spraying of Metasystox or Malathio @ 0.2% controls aphids.

Thrips (*Thrips maginis* and *Frankliniella tritici*)

Among insects, thrips are the most serious pests on rose. Thrips prefer dry hot conditions. Both nymphs and adult forms of this small insect suck the cell sap from the leaves, tender stems and flowers. Nair et al (1991) recommended the application of dimethoate and monocrotophos @0.05-0.1% to control thrips.

Red Scale (*Lindingaspis rossi* and *Aonidiella aurantii*)

Red scale insects attack the lower part of old stems of the rose plant and cover them with reddish brown waxy incrustation. The scale insect sucks the plant sap and the infested stems gradually dries up. Rainy season is the main period for infestation of red scale insects on rose plant. These insects can be removed by rubbing the infested stems with old soft tooth brush. These can be controlled by spraying Monocrotophos @ 0.1%.

DISEASES

Dieback (*Diplodia rosarum*)

Fungi such as *Colletotrichum* sp. have been found associated with dieback disease. The severity of this disease is noticed after the pruning operation. As its name indicates, symptoms of this disease are blackening at the tip of the stem, which rapidly spreads downwards. Blackening of twigs results from browning of internal tissues. The disease can be prevented by taking precaution at the time of pruning. The pruning cut should be slanting and made with sharp secateurs. Some systemic fungicides such as Bavistin 0.2% should be sprayed immediately after pruning. Removal of affected parts including about 2-3 cm below in the healthy part and use of Bavistin paste afterwards controls the disease.

Powdery Mildew (*Sphaerotheca pannosa* var. *rosae*)

In powdery mildew, initialy whitish tiny spots appear on young leaves. The entire plant displays white powdery appearance in severe cases, consequently reducing the vigour of the plant. Infected buds do not open and in open flowers the infection leads to discolouration. The disease can be controlled by spraying Karathane (0.1%) or Benlate (0.1%).

Downy Mildew (*Peronospora spersa*)

Downy mildew is a very common disease of roses and occurs under moist cloudy conditions. All species of cultivated and wild rose seem to be susceptible. Under cool and moist spring conditions, young leaves, stems and flowers may manifest purple to red or brown irregular spots. As the disease advances, lesions on leaves become angular and black with the possible appearance of white mycelium on the underside of the leaf. Advanced infections display yellowing of leaves with brown necrotic areas and noticeable leaf abscission. The disease can be controlled by spraying Ridomil or Dithane M-45 0.2%.

Black Spot (*Diplocarpon rosae*)

This fungal disease was first reported by R.D. Ansted (Bardoli and Ganguli, 1963) which causes major problems in cold climate of temperate region. The infested leaves show black spots on upper and lower sides. At later stages, leaves become yellow and fall off. The disease can be controlled by spraying Dithane M-45 0.15%.

Botrytis (*Botrytis cinerea*)

In Botrytis, the infection appears as brown patches on the flower buds. It is more common in temperate humid climates. In later stages, the buds become brownish and rot. Spraying of Bavistin 0.2% controls botrytis. Bar et al (2001) observed that susceptibility to *Botrytis* decreases with increase in concentration of calcium in the flower.

HARVESTING OF FLOWERS

The beauty of roses is best enjoyed when they are on the plants in the garden. Sometimes it may be necessary to have a few rose flowers for decorating the drawing room or the dining table. The flowers are harvested when they are at tight bud stage and when petals have just started unfolding. The flowers should be harvested either in the morning or evening by a sharp secateur. After harvesting, flowers should be dipped in clean water under proper ventilation. They should remain so for about one and half hour before any further operation.

REFERENCES

1. Balakrishnamuethy G and Roa VNM 1988 Studies on rooting of stem cuttings of Edward rose (*Rosa bourboniana*), *Madras Agricultural Journal* **75** 139-42.
2. Bar TA, Baas R, Ganmore NR, Dik A, Marissen N, Silber A, Davidov S, Hazan A, Krishner and Elad Y 2001 Rose flower production and quality as affected by Ca concentration in the petal, *Agronomi* **21** 393-402
3. Bhat M-S 1992 Micro propagation in roses. *Indian Horticulture* **37** 17-19
4. Bhattacharjee SK 1995 Cultural requirement of rose in *Advances in Horticulture* (eds) KL Chadha and SK Bhattacharjee, New Delhi: Malhotra Publishing House, 603-640 Pages
5. Bhattacharjee SK 1991 Influence of different months on flowering and flower quality of Rosa cv "Super Star", 29-31. *XIth All India Rose Convention*, Bengal Rose Society, Calcutta, India
6. Bhujbal B-G and Kale P-N 1973 Effect of some growth regulatore on rooting of rose. *Punjab Horticulture Journal* **13** 50-53
7. Biswas TD 1984 *Rose Growing-Principles and Practices*, New Delhi: Associated Publishing Co.
8. Bordoli DN and Ganguli D 1963 Black spot of roses caused by *Diplocarpon rosae* Wolf, *Indian Phytopath* **16** 255-59
9. Bose TK and Mukherjee D 1977 *Gardening in India*. Calcutta, Bombay, New Delhi: Oxford and IBH Publishing Co.
10. Malik RS and Dadlani NK 1984 Rose cultivation in India, *Indian Horticulture* **29** 27-30
11. Nair V-V, Raghunath P and Visalakshi A 1991 Control of thrips *Scirtothrips dorsalis* Hood on rose. Entomon **16** 327-329.
12. Nijjar GS and Rehalia AS 1977 Effect of nitrogen, potassium and phosphorus on growth and flowering of rose cvs. Super Star, *Indian Journal of Horticulture* **34** 75-79
13. Randhawa GS and Mukhopadhyay A 1986 *Floriculture in India*, New Delhi, Bombay, Calcutta, Madras: Allied Publishers Pvt. Ltd., 403-24 pages
14. Rout GR, Debata BK and Das P 1989 Micropropagation of *Rosa hybrid* cv Queen Elizabeth through *in vitro* culture of axillary buds, *Orrisa Journal of Horticulture* **17** 1-9

15. Singh AP, Prasad KV and Choudhary ML 2003 Cultivation of Rose in *Panorama of Rose Research,* New Delhi: All India Coordinated Project on Floriculture Division and Landscaping, Indian Agricultural Research Institute, 112-13 pages

16. Singh MK and Kumar Sanjay 2010 Effect of NPK on growth flowering of cut flower rose (*Rosa hybrida*) under polyhouse conditions, *Environment and Ecology* **28** 1498-1501

17. Uma S and Gowda JVN 1987 Studies on the effect of pruning, nutrients and their interaction on growth and flowering of rose cv Super Star, *Mysore Journal of Agricultural Sciences* **21** 455-60

18. Venkataraman KT, Singh B and Dadlani NK 1990 Seed set and seed germination studies on rose, *Indian Rose Annual* **8** 45-59

19. Vijaya N and Satyanartayan G 1990 Effect of culture media and growth regulators on *in vitro* propagation of rose. *In Horticulture New Techniques and Applications. Proc. Intn. Semi. On New Frontiers in Horticulture*, Bangalore, India

20. Yadav LP, Bose TK and Mukhopadhyay TP 1985 Influence of nitrogen and phosphorus on growth and flowering of roses, *Indian Rose Annuals* **4** 172-78

21. Zieslin N, Mor Y, Bachrach A, Haaze H and Kofranek AM 1976 Controlling the growth and development of rose plants after planting, *Scientia Horticulture* **4** 63-72.

CHRYSANTHEMUM

D. Mukherjee

INTRODUCTION

Chrysanthemum (Chrysos – golden; anthos – flower) is a herbaceous perennial extensively grown as an annual all over the world for its very attractive flowers with excellent vase life. In the international flower market, chrysanthemum ranks second among the elite group of flowers and is next to rose in demand. In many countries like Japan and USA, chrysanthemum is regarded as the number one flower. This popularity is due to the diversity in shape, size and colour of the flower and also for their excellent vase life. These plants are cultivated as pot plants for beautifying indoors and outdoors, grown in borders and beds, as cut flower for making bouquets and vase decoration and also as loose flower for making garlands and for religious purposes.

The earliest mention of chrysanthemum as 'Yellow Glory' is found in the writings of Confucius, the great Chinese philosopher, as early as 550 B.C. The seeds of chrysanthemum reached Japan through Korea as early as 386 A.D. and the first white flowering chrysanthemum was raised during 452-532 A.D. (Cuming 1964, Bose et al 2002). In India, the name of chrysanthemum plant was first recorded in Marathi literature in 13th Century A.D. by Sant Dnyaneshwar. However, the large flowering, standard types were introduced in India by the Britishers during 19th Century A.D.

During these two thousand and five hundred years of cultivation, thousands of cultivars have been developed in different parts of the world. Each year, new cultivars are being released for cut flower production, pot culture and for garden decoration. The present day, very attractive, and colourful chrysanthemums, of both the large and small types, are the result of extensive hybridization and spontaneously-induced mutations carried out by the chrysanthemum breeders of Japan, U.K., U.S.A., New Zealand, Australia, India and other countries. As a consequence, about 20,000 varieties of chrysanthemums have been reported from different parts of the world. A few hundred varieties have been released from India also. All these cultivars exhibit great variation in growth habit, flower form, a wide range of colours (white, cream, yellow, bronze, pink, red etc.) and an astonishing array of colour combinations and petal arrangements (incurving, incurved, reflex, irregular, decorative, anemone, pompon, ball type, Korean, spoon, spider, quill etc). The small flowered cultivars with floppy blooms are ideal for floral ornaments, garland making and hair decoration. The central disc florets may be numerous to nearly absent. The outer ray florets may have only one or two rows or may be numerous. Another group of chrysanthemum, the pot mums, may be dwarf, medium or tall.

A special characteristic of chrysanthemum plant is its amenability to be transmuted into the desired shape with synchronous flowering to express the skill of the flower grower. During the last few decades, with the advancement of knowledge and technology, manipulation of photoperiod (duration of light) and ambient temperature, and deployment of environment-controlled greenhouses, chrysanthemums can be grown throughout the year to get year round production of flowers (Machin and Scopes 1978).

ORIGIN AND CLASSIFICATION

Chrysanthemum (*Chrysanthemum morifolium* Ramat. or *Dendranthema* x *grandiflorum* Kitam) belongs to the family Asteraceae. It has recently been suggested that the plants should be known as *C. morifolium*. Chrysanthemum plants, comprising about 100-200 species, most probably originated in China. *Chrysanthemum morifolium* Ramat., the florists' chrysanthemum, is a very complex hybrid developed through interspecific crosses involving *C. indicum, C. japonense, C. Makinoi, C. ornatum, C. boreale, C. sinense, C. satsumense* and other species (Bailey and Bailey 1976).

Chrysanthemum flowers have been classified in different countries based on the number, size, form and arrangement of disc and ray florets. However, the classification made by the National Chrysanthemum Society of America has been widely adopted.

Single: It resembles a daisy flower with a flat central disc and one row of ray florets.

Semi-double: It has a flat central disk with two or more rows of ray florets.

Korean: It is very similar to single or semi-double type.

Anemone: Flowers have raised cushion-like disc florets with short rounded ray florets.

Pompon: Flowers are globular in shape. Ray florets are smooth, broad and incurved.

Decorative:	Ray florets are short, reflexed while the upper florets may tend to incurve. Blooms are more flattened than globular. Flowers are medium to large in size.
Incurved:	Ray florets are broad, curved upward and inward.
Incurving:	Ray florets are less incurved.
Reflex:	Ray florets are curved outward and downward.
Spider:	Tubular and elongated ray florets that are usually curved. Tips of ray florets are coiled or hooked.
Spoon:	Ray florets are tubular, straight with a spoon like flattened tip.
Quill:	Ray florets are tubular and elongated with tips open but not flattened. Flowers are double.

Chrysanthemum plants are also classified based on their responses towards temperature and photoperiod. Chrysanthemums are classified as 'standard' or 'spray' type. In the standard type, only the terminal flower bud is retained and all other flower buds are removed to nourish the terminal bud properly and to get a large terminal flower. In case of spray type, the situation is reversed; the terminal bud is removed at the earliest to encourage the lateral buds to bloom.

FACTORS INFLUENCING PLANT GROWTH AND FLOWERING

Extensive interspecific and intraspecific crosses have made the genetic make up of present day chrysanthemums very complex. This genotypic variations of chrysanthemum plants, combined with environmental factors, applied plant growth regulators, edaphic factors, and cultivation practices markedly influence the growth habit and flowering of chrysanthemums.

Genotypic variations

While a particular cultivar of chrysanthemum flowers at a particular time or season, there are marked variations in the time of flowering of different cultivars. Under natural conditions some varieties bloom in summer, some in the rainy season, some in autumn, while others in winter or in spring. The height of present day chrysanthemums vary from a few metres to a few centimetres and the flower size varies between 2.0-20.0 cm. Breeders are trying hard to develop varying types of chrysanthemums to suit the requirements of different flower growers. The performance of a chrysanthemum plant reflects the cultivar and its environment interaction.

External Factors

Light

Among the different environmental factors, the influence of light is more pronounced than that of temperature, humidity or CO_2. Chrysanthemum is a short day crop; the plants form flower buds and start flowering if exposed to short days. Light affects the growth of chrysanthemum plants by influencing the plant growth and leaf production. For flower initiation, the minimum length of continuous dark period every day has been found to be more than 9.5 hours. It has also been observed that a well-grown plant responds to short days more rapidly than a sick plant. On the other hand, higher light intensity during the day time induces early flowering in short days. In general, chrysanthemum plants require high light intensity. Plants grown under low light intensity become tall with thin stems, larger leaves and inferior quality flowers. Under white fluorescent tube lights, the internodes of chrysanthemum plants become shorter, and the leaves become smaller than the plant when grown under natural light. Also, the light from incandescent lamps increases the internode length.

If short day conditions are created by keeping the plants in a dark chamber (polyhouse with black polythene) for 14 hours a day (from 5 pm-7 am) or more, flowering time can be hastened by up to one month for many cultivars. Similarly, blooming can be delayed appreciably by exposing the plants to long days i.e. illuminating the plants with a combination of fluorescent tubes (40 W) and incandescent lamps (60 W) by hanging them alternately, 1.2 m above the plants, so that the plants do not get more than 9 hours of continuous dark period. Illumination during the middle of the night has been found to be more effective.

Growers force the chrysanthemum plants to flower by providing shade with the help of black cloth from 5 pm - 8 am. The black cloth creates an environment of 9 hours day and 15 hours night. The varieties, which usually flower in November under natural conditions, start flowering from September onwards. By controlling the photoperiod and temperature, one can have chrysanthemum flowers for 365 days a year. To have a proper balance between vegetative and reproductive growth, a few weeks of long day conditions, immediately after planting the cutting, is essential. A crop should be provided with two or three weeks of long days for vegetative growth followed by 10 weeks of short days for the initiation of flowering and their opening. However, the short day requirement of chrysanthemum varies from variety to variety and it may vary from 7-15 weeks. Pompons usually require 13-14 weeks of short days. For flower initiation, long nights and a temperature of around 16°C are essential. This temperature requirement varies from variety to variety i.e. from 13-17°C. Hence, a temperature regime of 16-17°C should be maintained till the flower buds are formed. Low light integrals and temperatures above 20°C can retard the flower growth. Additionaly, spectral quality of light has also been found to influence the photoperiodic response in chrysanthemum. A wavelength between 600 and 680 nm (red) was found to be most effective for initiation of flowering in short day plants.

Temperature

Based on the response of chrysanthemum plants to temperature, Cathey (1954) classified the chrysanthemums into three groups – thermopositive (flowering is inhibited by low temperature), thermonegative (flowering inhibited by high temperature) and thermozero (flowering not influenced by either high or low temperature). In the thermonegative group, the flower development is encouraged by increase in temperature upto 16°C but the process of flower development is affected when the temperature reaches 21°C or more. The flowering of thermozero cultivars is relatively unaffected by temperature during the long day period. The flower development process in thermopositive cultivars is particularly delayed at 10°C if the temperature during long day is also maintained at 10°C (Cockshull 1985). Most of the present day chrysanthemum cultivars, which are cultivated for round-the-year flowering, belong to thermozero group as they have the ability to grow and flower under a wide range of temperatures. However, the interaction between light and temperature is more important than the individual effect of light or temperature.

A temperature regime of 21-23°C during the day and 18-19°C during the night is congenial for chrysanthemum plants. The ventilators of the greenhouses should be opened if the temperature inside the greenhouse exceeds 23°C. If the temperature exceeds 35°C, there is every possibility that the bud formation will be retarded or in some varieties the bud formation will be completely stopped. As night temperature determines the bud formation, care should be taken that the temperature is not below 17°C as it adversely affects flower bud formation in many varieties. Hence, for most of the chrysanthemum cultivars for cut flower and pot plant production, the growers usually maintain 18-19°C during the night till the flower buds develop.

For balanced root and shoot growth, the optimum air and soil temperatures are found to be 16°C and 18-21°C respectively. The effect of night temperature is found to be more important than day temperature in regulating growth and flowering in chrysanthemum. Taller plants with thicker stems and larger flowers are obtained when the plants are grown at a night temperature of 10°C or more for 9-10 hours and the remaining part of the day at 16°C or more. Inappropriate temperature during short days will delay floral initiation and development. Temperature has a profound influence on flower initiation and early flower bud development during the first few weeks of short days. It has been recommended that the night temperature should be maintained at 16°C from flower initiation to development of visible flower buds and the temperature should be lowered to 10-13°C thereafter until they are harvested. It has been noted that 8-12 hours day length produces acceptable flowering at 18°C while 8-10 hour day length is required when the temperature is 24°C. High temperatures may increase the number of florets formed on receptacle but the bloom may get damaged. Flower colour, especially pink and bronze, fades at high temperature. Highest percentage of flowering and higher dry weight are recorded when the plants are grown at 20°C throughout.

Relative Humidity

No negative consequence of high humidity was observed, but there was a strong increase in the dry weight of all plant organs. In high humidity, the dry weight of the total plant is increased upto 39% higher than in the regular regime (Korner and Challa 2004).

Carbon dioxide (CO_2)

Increased CO_2 in the growing environment results in greater stem length, fresh weight of stems and flowers, early flowering and increased yield of flowers. In general, CO_2 concentration is more inside the greenhouse as the CO_2 evolved during the night is trapped within. With the rising of the morning sun, plants growing inside the greenhouse utilise this CO_2 for photosynthesis and more photosynthates are produced. As a result, the plants grown in the greenhouse are much better in health and flower production. However, the response of plants to carbon dioxide enrichment depends on many factors such as age of plants, time of treatment, cultivar, availability of light and moisture.

Presently, commercial greenhouse growers use 1000-1500 ppm of CO_2 that starts one hour before sun rise and continue upto one hour before sunset. Flowering is advanced by a week. The flowers are larger and are borne on stronger stems when compared with the plants not given the supplementary CO_2.

APPLICATION OF PLANT GROWTH REGULATORS

Growth regulating chemicals have been found to appreciably influence plant growth and flowering of chrysanthemum. Among the different growth regulators, the efficacy of GA_3 is very pronounced. Application of GA_3 has been reported to increase the plant height, supplement the low temperature requirement for promoting flowering, advance flowering and promote rooting of cuttings. Many workers have also reported that application of GA_3 improves flower quality, size of flowers, duration of flowering, flower yield and vase life of flowers. Use of root promoting hormones such as IBA and NAA increases the rooting percentages of the chrysanthemum cuttings. Ethephon increases branching in chrysanthemum. In combination with GA_3, application of ethephon has been found to delay flowering. Multiple applications of ethephon have been found to increase plant height, delay flowering and increase fresh weight of cut flowers. BA (benzyladenine) significantly increases branching of the chrysanthemum plants, whereas application of triacontanol increases the number of inflorescences and improves the quality of flowers.

On the other hand, a number of growth retarding chemicals such as alar, chlormequat, phosfon, paclobutrazol and uniconazole have also been extensively tried on chrysanthemum. Many of these growth retarding chemicals are being used in commercial floriculture to produce bushy, well-grown chrysanthemum plants with attractive appearance and increased number of flowers. Among all growth retarding chemicals, B-Nine or Alar (daminozide) is extensively used in chrysanthemum culture. It has been reported that alongwith Alar, GA_3 can counteract the adverse effect of high salinity and produce vigorous plants with more number of flowers as compared to untreated plants. The growth inhibitors such as maleic hydrazide (MH) and TIBA have also been tried on chrysanthemum plants. MH has been found to reduce plant height, increase branching and produce more flowers per plant. Application of MH alongwith NAA has been reported to delay flowering. A number of other growth retarding chemicals such as Amo-1618, ancymidol (A-Rest), flurmidol and EI-531 (cyclopropyl-a-4-methoxyphenyl-5-pyrimidinemethanol) have also been tried on chrysanthemum plants with varying degrees of success.

PROPAGATION

Chrysanthemum plants are propagated both by sexual and asexual methods or by vegetative methods. The former method is used by the breeders to develop hybrids, while the latter method is adopted by the amateurs and commercial growers to maintain the purity of the cultivar. Micro-propagation is practised for getting disease-free plant materials on a very large scale. Now, with the advancement of knowledge and techniques, somatic embryogenesis and transgenic chrysanthemums have also become a reality.

Seed Propagation

Cultivars which bloom with conspicuous disc usually produce plenty of seeds. Poorly nourished plants are also known to produce seeds better than the well-nourished plants. The seeds should be collected after the blooms are completely dry. The small seeds require care in handling before and after sowing. Fine compost is used to cover the seeds after sowing and watering so as not to expose the seeds to the sunlight. Seeds germinate within a few days after sowing. Single Korean and charm types are usually raised through seeds.

Vegetative Propagation

Suckers

This is an old method for propagation of chrysanthemum and is still practiced by some growers. The plants propagated through suckers are tall and lanky, and the flowers are comparatively smaller in size as compared to the plants raised through cuttings. The suckers are separated from the mother plants when they have 5-6 green leaves. Before planting the suckers, the long roots should be trimmed to 1.5-2.0 cm length and a few basal leaves should be removed to avoid direct contact of leaves with soil. Rooted suckers are planted in raised beds or small pots in shade to encourage profuse rooting. Rooted cuttings obtained from suckers exhibit more vigorous vegetative growth but flowering is delayed. Suckers are better propagating material than the lateral shoots (Sun et al 1999).

Cuttings

The commercial growers propagate chrysanthemum plants through terminal stem cuttings. The modern method of propagation of chrysanthemum consists of first raising of disease-free mother plants and then growing them under controlled environment. Cuttings are taken from these specially raised mother plants. It has been observed that large number of cuttings are obtained when the mother or stock plants are grown at 21-27°C than those grown at 16°C. For nutrition of mother stock plants, the ratio of N and K_2O should be 3:1. Phosphorus also plays an important role in rooting of cuttings. These days cuttings are collected by snapping the shoot tips from stock plants by hand to avoid any possible contamination of virus diseases through cutting tools. If the stock plants are subjected to long days and fed with a weak nitrogenous liquid fertilizer, the shoots snap easily from the mother plants. If optimum conditions for rooting are provided, there is no difference in rooting performance between a nodal cutting and an internodal cutting. However, for commercial purposes, cuttings are propagated from new terminal shoots of the mother plants. In horticulturally developed countries, uniform cuttings are selected taking the help of computerized image analyser. An ideal cutting is 10-12 cm long with a basal stem diameter of 5-6 mm.

The rooting medium should have good drainage, should be light weight and should be free from any toxic substance. Sand, vermiculite, perlite, peat, sphagnum peat moss etc. alone or in combination with other media can be effectively used for rooting of chrysanthemum cuttings. A commonly used rooting medium is a mixture of 60% fine moss peat and 40% perlite. A little quantity of lime should be incorporated into the rooting medium. The pH of 6 should be maintained around. The rooting medium should be atleast 12-15 cm deep. Use of root promoting hormones like IBA and NAA are found to be very effective in increasing root formation of cuttings. Callus formation and rooting increased with the increase in temperature up to a certain limit and a night temperature of about 18-19°C has been found to be optimum for rooting of chrysanthemums. Mist sprays should be used intermittently during the day time until rooting is accomplished. During sunny days, the cuttings may be provided with mist sprays 3 times an hour for a period of 5-10 seconds each time. As soon as the cuttings become ready to absorb moisture from the medium, misting should gradually be discontinued. Care should be taken to switch off the mist well before sunset i.e. around 3-4 pm, so that the leaves become dry before the sunset to minimize the fungal infections. Substrate temperature between 18°C and 21°C is ideal for root formation of cuttings. The pace of root formation decreases rapidly at a temperature below 16°C. Supplementary lighting has been found to increase the number of cuttings per plant and allow better rooting. The roots appear within 10-12 days and the cuttings are ready for planting within 2-3 weeks. During the summer, cuttings should be protected from direct sunlight by using shade net.

Chrysanthemums can also be propagated through leaf cuttings, layering etc.

Since chrysanthemums are commercially propagated through cuttings, pre-rooting storage of cuttings in the dark is a common practice among growers and nurserymen.

Micro-propagation

Chrysanthemum can be multiplied at a very fast rate by culturing different explants under *in vitro* condition. For *in vitro* culture, different types of explants such as shoot tips, nodes, leaves, peduncles, flower petals etc. have been tried with varying degrees of success. Clonal propagation through *in vitro* culture resulted in production of disease-free plantlets in a very short time. Plants raised through *in vitro* culture showed higher yield and better quality flowers.

Before initiating chrysanthemum cultivation, one should take into consideration the requirement of the target customers or the market preference. For example, for worshipping purpose or for making garlands, the large flowering types are not suitable. One should grow the small flowering or the Korean types. On the other hand, for making bouquets or for cut flower decoration, the large and medium sized flowers are preferred. Different countries have special fascinations or preferences for a particular type. Decorative chrysanthemums have a very good demand in U.K., while Germany and Netherlands have a preference for spiders. The spray chrysanthemums have a very good demand in Sweden, Finland and Denmark.

The success of chrysanthemum cultivation, irrespective of whether done in open field or greenhouse, to a great extent depends on selection of cultivars. In areas with high light intensity, cultivars which take 7-9 weeks time to flower should be selected for greenhouse cultivation, whereas in areas with low light intensity, the long duration cultivars which require 12-14 weeks time to flower should be selected for production of quality flowers. However, the cultivars belonging to 10 weeks response group can be successfully grown in both the light conditions. Some of the year-round-flowering cultivars belong to this group. Many standard types are 10 week cultivars.

CULTIVATION

Open Air Cultivation

Large and spray type chrysanthemums are grown for cut flower production and for pot culture while the small flowering types are cultivated for loose flowers for making garlands and worship. In plants grown under open field conditions, the yield is low and the quality of flower is much inferior as compared to the greenhouse grown plants.

Soil

Chrysanthemum has a shallow and fibrous root system. Hence, sandy loam soil is ideal for growing chrysanthemum. The land should be thoroughly prepared by repeated ploughing and spading to a depth of 40-50 cm. A regular supply of calcium is essential for proper growth and flowering in chrysanthemum. Liming should be done at the time of bed preparation. Care should be taken not to apply lime and FYM together. The FYM should be incorporated with the soil at least 3-4 weeks after the application of lime. Incorporation of lime supplies calcium to the soil, makes the soil loose and increases the pH of the soil. Addition of FYM increases the organic matter content and the water holding capacity of the soil. Chrysanthemum prefers a soil with a pH of around 6.5.

Soil Sterilization

To have a healthy crop of chrysanthemum, the soil should be sterilized properly to reduce the population of harmful pathogens. The soil can be sterilized through solarization or by use of steam or chemically by using Methyl Bromide @ 100 g per square metre or Formaldehyde @ 2%. After the chemical treatment, the soil should be covered with polythene for 72 hours. Planting of cuttings can be started a week after removal of the soil cover.

Drainage

Chrysanthemum is very sensitive to water logging. Raised beds facilitate good drainage and are ideal to:

(i) Maintain a constant level of ground water

(ii) Remove the excess salt

(iii) Maintain the soil structure in good form

Sub-surface drains should be installed 80 cm to 1 m below the ground level. The pipes for the drainage may be installed 3 m apart.

Water

The availability of good quality water is essential for chrysanthemum cultivation. Good quality water plays a more vital role than the soil type. Low salt (NaCl) concentration is essential for quality flower production.

Planting Time

To have a good crop of flowers, it is very essential that the plants complete their vegetative growth before the onset of natural short days. For proper flowering, a balance between the vegetative and the reproductive growth is necessary. In the hills, the chrysanthemum cuttings may be planted during June-July. The plants flower from mid-October to end of January under natural conditions. However, due to cold in open field conditions, the leaves dessicate and turn brown in colour during December-January rendering the plants unsuitable for marketing. In the plains, the planting time varies from place to place. However, the plants should initially get the exposure to long days for a few weeks for vegetative growth followed by short days under natural conditions. If the beds are not ready for plantation, the cuttings may be stored at a temperature of 1-4°C. The relative humidity should be maintained at about 90%. However, the storing period should be as short as possible.

Planting Density

Plant density varies according to cultivar, place and time of planting. In open cultivation, the plant density is usually less than the green-houses. The plant density will also vary with the types of chrysanthemum and cultivars. For large and medium sized flower cultivars, grown as standards, the plant density varies between 16-20 plants per square metre. The plants will produce 1-3 good quality cut flowers per plant. For spray and small flowered cultivars, the plant density is slightly higher and 20-30 plants per square metre are planted depending on the growth habit of the cultivar.

Nutrition

Chrysanthemum is a heavy feeder and has large requirements for nitrogen, phosphorus and potash. At the time of soil preparation, liming @ 10 g per square metre should be done atleast one month before the planting of the cuttings. At the time of planting the cuttings, the final bed preparation may be done by incorporating well decomposed FYM @ 5 kg per square metre and whole of the phosphorous and potash. The nitrogen should be applied in two split doses; 15 days and 30 days after planting of cuttings. At the early stage of plant growth, the emphasis should be on nitrogen. As the buds appear, the portion of potassium should be increased. Plants need phosphorus throughout the growing period and it is best applied as basal dressing, which will be available to the plants slowly. The amount of nitrogen should be reduced when the flower buds appear. The recommended dose is N (40 g), P_2O_5 (20 g) and K_2O (80 g) per square metre.

Nitrogen deficiency reduces plant growth and results in yellow green leaves of small size. Flowering is delayed, and the number and size of bloom is reduced. The effect of nitrogen also varies with the source of nutrient. Ammonium nitrate and calcium nitrate appear to promote flower production as compared to other nitrogenous fertilizer. Ammonium nitrate also produces plants with stronger stems.

The first symptom of phosphorus deficiency is manifested in reduction of size of the new leaves. The leaf margins become brown due to potash deficiency. A regular supply of calcium is essential for normal growth of chrysanthemum. Slight deficiency of calcium results in complete cessation of top growth and death of roots.

Application of liquid manure is beneficial for chrysanthemum. The liquid manure can be prepared by decomposing fresh cow-dung and oil cake, 1-2 kg each in 10 litres of water, in a drum for about 2-3 weeks. Dilute it with water to tea colour and apply @ 500 ml per plant. A fertilizer mixture containing 10 gm of potassium nitrate, 30 g of ammonium sulphate and 30 g of super phosphate dissolved in 10 litres of aforesaid liquid manure should be applied at an interval of 10-15 days. Liquid manure should be continued till all the buds are half open.

Watering

Chrysanthemum plants can tolerate appreciable water stress. Over-watering is very harmful to the plants. The plants need adequate water during active vegetative growth when new leaves are being formed. After the formation of flower buds, lesser amount of water is needed by the plants. Irrigation should be done at an interval of 2-7 days depending on the soil and the weather conditions.

Pinching

Removal of terminal growing portion of stem is known as pinching. Pinching operation reduces the plant height and promotes axillary branches. Time and severity of pinching depends on the type of chrysanthemum and the required results. The time of last pinch influences the blooming date and also controls the length of shoot and plant height.

Disbudding

The practice of removal of immature flower buds to either get a small number of large flowers or large number of small flowers is known as disbudding. In case of standard types, only the the main terminal flower bud and a few healthy flower buds are kept, and all other flower buds are removed to channelise the nutrients only to the selected buds. In case of spray types, the large apical bud is removed and the axillary buds are allowed to develop. Disbudding of the spray cultivars also varies with the type of spray produced. The correct time and method of disbudding operations are important factors in the production of high quality plants and flowers. Buds should be removed as soon as they can be easily rolled between the thumb and the forefinger.

Weed Control

In India, application of herbicides to control weeds of the chrysanthemum field is rare. In countries abroad, weedicides like alachlor, oxyfluorfen, nitrofen, oxidiazon, oryzalin etc. are found to be effective to control weeds. For manual removal of weeds, hand hoeing once a week or a fortnight, depending on the season and weed population, can keep the field weed free. This hoeing also helps in conservation of soil moisture.

Mulching

Mulching helps to conserve moisture in the soil and to check the weed population. For most desired parameters such as plant height, length of flower stalk, number of flower sprays or plants, flower diameter, longevity of flower, and duration of flowering, black plastic mulch gives significantly better performance compared to transparent mulch. Marked variations in response to mulching treatment among different cultivars are recorded. (Mukherjee and Raja Ram 1999a).

Cultivation in Pots

Pot culture of chrysanthemum is popular in areas where the flower markets are near by. This is applicable to all types of flowers. Standard types of chrysanthemums, which have large and medium flowers are usually grown in 15-20 cm pots. One pot should contain one plant, which will bear 1-3 flowers. The small flowering types are grown in bigger pots such as 25-30 cm pots. As these plants are bushy and produce hundreds of flowers per plant, they require bigger pots for proper performance. Rooted stem cuttings of large flowering standard types are planted during June-July. Initially the plants are planted in 10 cm pots, and are then gradually shifted to bigger pots i.e. from 10-12.5 cm, followed by 15 cm and ultimately to 20 cm pots to make the plants stocky. All side shoots are removed to obtain a single erect stem. Except the terminal bud, all other buds are removed if only one large flower is required. The plants are pinched once if more than one bloom is desired. The plants are staked, preferably with a bamboo stick, to keep the plants erect. In many countries, growth retarding chemicals such as Alar and Phosfonare routinely used to produce healthy and attractive stocky pot plants.

Suckers of small flowering varieties are planted during January-February. These plants are pinched several times to encourage branching and to have a large bush with hundreds of flowers per plant. The plants are provided with support by inserting a number of stakes (4-6) around the periphery, tilting outwards, and the stakes are held together with the help of a thread tied with the stakes, encircling the plants, and preventing the outer branches to fall.

Rooted cuttings of pot mums are grown in 8-12 cm pots. These plants attain a height of 10-15 cm and usually bear 30-50 small blooms. To have a vigorous bushy growth of these miniature pot mums, the plants are very frequently provided with liquid feeding.

Greenhouse Cultivation

Chrysanthemum flowers have year round demand. For flower production, it is essential to control the photoperiod and the temperature. As this is not possible under open field conditions in developed countries, chrysanthemum plants are extensively grown in greenhouses, for year round commercial production of pot plants and cut flowers. In recent years, chrysanthemum is also being grown in tropical and subtropical areas in greenhouses or polyhouses for production of quality flowers. In order to provide an environment favourable for growth and year round flowering, greenhouses are equipped with facilities to control temperature, light (intensity and duration), humidity and even to enrich the internal environment with carbon dioxide. Inside these environment-controlled greenhouses, the irrigation and fertilizer application systems (fertigation) are regulated by the computer controlled devices. Greenhouses are designed in such a way that maximum amount of sunlight enters the greenhouse. Glass, polycarbonate sheets, fibre glass reinforced plastic (FRP), UV stabilized polythene sheets etc. are used as cladding material for these greenhouses. Each of these cladding materials has its merits and demerits. Out of all these cladding materials, polythene sheets are the cheapest and easiest to install. However, the longevity of these sheets is only 2-3 years.

Earlier, the glass greenhouses had ventilators to regulate the inside temperature and humidity and to provide air circulation among the plants. With the advancement of technology, most of the greenhouses now are fitted with fan pad cooling system. By operating the exhaust fans, the hot air from inside the greenhouse is sucked out and outside air is sucked inside through the cool cell to cool down the greenhouse. As a result, the temperature inside the greenhouse can be lowered by 10-15°C. During winter, the temperature inside the greenhouse can be increased by using heaters. Many of these greenhouses are equipped with light control facilities, to provide long day or short day conditions to the plants growing inside the greenhouses. In the greenhouses, provision of lighting is made for proper growth of the greenhouse plants during the period of inadequate natural light and also to prevent flower bud formation. A minimum of 10 ft candle (100 lux) light is sufficient to prevent premature bud formation in chrysanthemum and this light can be arranged through 60 W incandescent bulbs with reflectors spaced 1.0 m apart and 1.0-1.2 m above the plants.

Shading is necessary for chrysanthemum during long days to induce flower bud formation. Short day condition is created by covering the plants with black cloth or black polythene sheet.

Carbon dioxide enrichment inside the greenhouse is done preferably in the morning by burning propane in burners. For production of photosynthates, humidity is also maintained between 70-90%.

Shading

For year round chrysanthemum flower production, when the natural day lengths are too long for flower bud initiation, short day conditions are artificially created by covering the plants with black cloth or black polythene. In floriculturally developed countries, this shading operation is carried out by mechanized automatic time switch controlled system at a very fast speed. This system also saves labour expenses. When the plant reaches a length of about 40 cm, the development of flower buds can be initiated. To initiate flower bud development, the plants should be provided with night length or short day conditions for 13 hours or more. The duration of this short day treatment, which is continued until the plants flower, varies between 7-11 weeks depending on the cultivars. Once the short day requirement of a particular variety is known, one can precisely calculate the starting date and duration of the short day treatment to get the chrysanthemum flowers.

Substrate

The substrate for chrysanthemum culture consists of various organic and inorganic material. A number of materials such as soil, peat, perlite, bark, bamboo compost, cocopeat and spent compost are used as a substrate for chrysanthemum cultivation with varying degrees of success. Loam soil with plenty of organic matter in combination with sand, well decomposed FYM, cocopeat or peat moss are used in benches for growing chrysanthemums. Soil, organic matter and sand are usually used in the proportion of 1:1:1 in order to keep the mixture friable and loose for adequte aeration and drainage. The pH of the medium is maintained around 6.5. The media should be sterilized by solar heating or by using steam or chemicals such as methylbromide and formaldehyde.

Planting Time

Flowering of spray chrysanthemum cultivars, in year round cultivation, depends on the date of planting.

Planting Density

The density of planting varies according to cultivar, growing environment and the time of planting. It has been observed that the best population is 48 cuttings per square metre as it results in normal development of plants and blooms, and also from the point of planting cost and market returns. Very close spacing can suppress the plant growth and affect the flower quality. As during the winter the light intensity is less as compared to summer and rains, a lesser plant density of 40 plants per square metre can give higher percentage of first grade flowers as compared to 48 or more plants per square metre. The number and average weight of flowers per spray also increases to a certain extent with wider spacing.

Planting of Cuttings

It is necessary to plant the cuttings soon after they are rooted. As already stated, a number of materials such as soil, peat, perlite, bark, bamboo compost, cocopeat and vermicompost have been used as a substrate for cultivation of chrysanthemum plants with varying degrees of success. Loamy soil rich in organic matter, in combination with sand, well decomposed FYM, cocopeat and soilrite can also be used. Rooted cuttings are graded according to size, and only healthy cuttings of uniform growth are planted. While planting, the cuttings should not be planted very deep and the roots should be spread out. Before planting, the soil or other substrates should be thoroughly sterilized or disinfected to minimize the infection of soil pathogens. Deep planting (more than 1.2 cm) or planting in too wet soil often results in the death of the plants and the young roots may suffer from lack of oxygen. Inside the greenhouse with temperature controlled facilities, cuttings can be planted at any time of the year.

Salinity

Chrysanthemum plants are very susceptible to high salinity of the growing medium. With increase in salinity, the vital plant nutrients become unavailable to the young plants.

Watering

Chrysanthemum roots are very sensitive to water logging but can tolerate water stress to a great extent. Water should be given for 3-5 minutes immediately after planting the rooted cuttings. During the following days, water should be given according to the weather and the moisture content of the soil. This system should be continued until the plants are well established. Thereafter, water should be given once or twice a week, depending on the requirements of the crop. Chrysanthemum plants need profuse quantity of water during active vegetative growth. Water requirement declines gradually after the formation of flower buds as no new leaves are formed. The best growth is observed in plants grown in soils that are constantly kept moist immediately after planting till the plants are properly established. Frequent watering during the active period of growth is required for proper maintenance of the foliage and the flower quality.

Different systems of irrigation are practised in different countries of the world. The overhead systems such as mist and sprinklers should be used before the appearance of the flower buds and then should be switched over to drip system in order to save the flower buds. Moreover, with the overhead watering system, there is a danger of spreading fungal diseases. With sub-irrigation systems, the problem of wetting the foliage can be avoided, but clogging of the drippers may occur. With trickle irrigation and other similar systems, it is economically possible to maintain adequate moisture with good aeration of soil. By the use of trickle irrigation, a substantial reduction in the consumption of water is possible without affecting the yield and the quality of cut flowers. While using the constant level method, the problem of anaerobic conditions and high salt concentrations is likely to be developed in growing media.

Nutrition

In chrysanthemum, the total uptake of major nutrients was reported to be 20 g N, 10 g P_2O_5 and 40 g K_2O per square metre. The nutrient requirements at different stages of growth revealed that all the basic nutrient requirements (N, P and K) were high during the rapid growth period. Nitrogen requirement is high, particularly, during the early stages of growth. In the commercial production of flowers, maintainance of high level of N during first 6-7 weeks of plant growth is very essential and a moderate deficiency of N at this stage can not be compensated by N fertilization at subsequent stages. Flowering is delayed and size of blooms and their numbers are reduced with N deficiency. However, excess N adversely affects the flower quality and longevity.

Phosphorus is essential for the chrysanthemum plants throughout the period of growth and development, and is especially important during the time of flower initiation and subsequent development of flower buds. It promotes stem length and flower size. In case of phosphorus deficiency, the size of new leaves is reduced but their colour remains green. Phosphorus deficiency delays the flowering and decreases the number and size of flowers. Hence, phosphorus should be applied best as basal dressing so that it becomes available to the plants throughout their growing period. A positive correlation between flower production, and N and P contents of soil has been observed.

In chrysanthemum plants, potassium requirement increases with plant growth and remains high after bud formation till flowering. Deficiency of potash causes delayed flowering with dull colour and it also adversely affects the keeping quality of flowers. However, excess of potash results in suppressed plant growth, dark green leaves and may result in Mg and Ca deficiency.

A concentration between 200-300 ppm of N, P and K has been found to be optimum and should be used through irrigation water during fertigation of chrysanthemum plants under greenhouse conditions. Foliar application of nutrients is also an efficient method because higher percentages of nutrients are absorbed through leaves as compared to other methods of feeding. Micronutrients are mostly applied through foliar sprays.

Pinching

The pinching operation plays a very significant role in making the chrysanthemum plants bushy with increased number of branches and flowers. The time and severity of pinching varies with the variety. The pinching operation reduces the plant height, delays the flowering, increases the flower yield, and gives a uniform appearance to the flower crop.

Disbudding

As stated earlier, disbudding is the practice of removal of unwanted immature flower buds either to get a small number of large flowers or to get a large number of small flowers. Depending on the situation, either the main terminal bud is retained and all other buds are removed or the terminal flower bud is removed to promote the development of the axillary buds as in the case of spray types. If disbudding is done too early, damage may be caused to the side buds. Hence, disbudding should be carried out 2 or 3 weeks before harvesting of flowers. Carvalho et al (2002) reported that when no lateral flower buds were removed (control), higher assimilate supply resulted in more flowers rather than in larger flowers in chrysanthemum cv. Reagan Improved.

Mulching

Mulching shows beneficial effects on the yield of chrysanthemum flowers. A number of mulching materials such as reflective, clear, black or green films or plastic sheets have been tried.

PESTS AND DISEASES

Chrysanthemum plants and flowers are much susceptible to the attack of different types of insects, pests and diseases. All these insects, pests and diseases cause damage to the plants and flowers to varying degrees.

Insects and Pests

Aphids, thrips, red spider mites, caterpillars, white flies, leaf miner, mealy bugs, grubs, beetles and nematodes infest the chrysanthemum plants in different seasons of the year and sometimes cause extensive economic loss to the growers.

Aphids (Chrysanthemum aphid-*Macrosiphoniella sanborni*, cotton aphid-*Aphis gossypii*, green peach aphid-*Myzus persicae*, peach leaf curling aphid-*Brachycaudus helichrysi).*

The chrysanthemum aphids are large in size and dark chocolate brown in colour. The peach aphids, which feed on leaves are green in colour, rarely reddish also. These aphids settle on the young terminal shoots and on the under surfaces of the leaves. They suck the plant sap. Severe infestation by these tiny greenish brown pests result in curling of leaves, stunted growth, and gradual drying and death of plants, especially the young plants. Moreover, the aphids also transmit a number of viral diseases.

In order to minimize the aphid attack, the weeds growing within and around the plant production area should be removed. Many organophosphate and carbamate insecticides are no longer effective against the aphids, whereas the pyrethroid insecticides may encourage the aphid reproduction process. Application of endosulfan (Thiodan) @ 2 mllitre or malathion (Malathion) @ 2 ml per litre or methyl parathion (Metacid-50) @ 2 ml per litre as foliar spray can control the aphids. Care should be taken not to use dimethoate (Rogor 30 EC) on chrysanthemum plants as this chemical causes phytotoxicity to the chrysanthemum plants. A number of other plant protection chemicals such as acephate (Orthene or Pinpoint), imidac-loprid (Marathon IG), oxamyl (Oxamyl 10 G), chlorpyrifos (Dursban) and azadirachtin (Azatin) are found to be highly effective against aphids. Some growers also use Pirimor and Hostaquick or Thiodan and Hostaquick against the aphids.

For biocontrol of aphids, predatory midge *Aphidolotes amphidimyza* and lacewings *chrysopa* species have been successfully used. The fungi *Beauveria bassiana* and *Verticillium lecanii* have also been found to be effective for control of aphids.

Thrips (Onion thrips - *Thrips tabaci*, western flower thrips-*Frankliniella occidentalis, Thrips palmi, Caliothrips phaseoli*).

Different types of thrips infest the chrysanthemum plants. Thrips feed on the growing points causing mottling and distortion of leaves. They also impart silver colouration due to separation of upper and epidermal tissue from the rest of the leaf. The insects also damage flowers of summer blooming cultivars. The affected flowers ultimately dry out. In recent times, the western flower thrips have been reported to cause substantial damage to the chrysanthemum crop. The thrips population becomes less virulent under high light intensity. On the other hand, the damage is more during winter or under shade. Crops receiving high nitrogen fertility and irrigation suffer less from thrips attack.

To control the thrips, malathion (Malathion) @ 2 ml per litre or monocrotophos (Nuvacron) @ 2 ml per litre should be sprayed on the plants. Soil application of aldicarb (Temik) @ 5 g per square metre can reduce the thrips infestation. Use of chlorpyrifos (Dursban), bendiocarb (Dycarb), azadirachtin (Azatin) and acephate (Orthene) are also found to be highly effective against thrips.

For successful control management of thrips, continuous monitoring of the thrips movement is essential. Many growers and pest specialists install sticky traps inside the greenhouse to monitor the thrips movement. Usually yellow coloured sticky traps alongwith blue or white traps are used for this purpose. Thrips are normally seen on traps before they become visible on plants. However, one should not depend on traps alone. Regular inspection of plants should be carried out. Use of aluminized reflective tapes and aluminized shade cloth or screens reduce the number of western flower thrips movement from the fields to the commercial greenhouses.

For biological control of thrips, predatory insects such as *Orius insidiosus* and *O. tristicolor* are found to be very effective against western flower thrips and onion thrips. *Weoseiulus cucumeris* and *N. barkeri* are also found to be effective against thrips. The entomopathogenic fungi like *Beauveria bassiana* are reported to be an excellent biocontrol agent to control thrips.

Red Spider Mites (*Tetranychus urticae*)

These are small insects that are generally found on the under surface of the leaves, but in case of severe attack the mites may be observed all over the leaf surface. The adults are scarlet red in colour. Mites puncture the leaf tissue and suck the plant sap. Along with the plant sap, the removal of chlorophyll and other plant pigments result in the characteristic blotching of leaves. They also cause damage to the opening buds.

High nitrogen is associated with severe mite infestation. Spider mites prefer moisture-stressed plants. Proper watering to the plants can control the mite population. The area near the chrysanthemum crop should be kept weed-free as far as possible. Frequent applications of pesticides are often necessary to keep the population of mites under control. To control red spider mites, foliar sprays of methyl demeton (Metasystox) @ 1 ml per litre of water or dicofol (Kelthane) @ 2 g per litre are recommended. It is advisable to make atleast two or three applications of the pesticide at an interval of 5-7 days. However, due to long residual effect, 1 or 2 sprays of pyridaben can probably control the red spider mites. To control the spider mites, granular pesticides such as oxamyl (Oxamyl 10 G), disulfoton (Di-Syston) and other pesticides such as dichlorvos (Fulex DDVP Smoke), dichlorvos (GH-19), and sulfotepp (Plantfume 103 Smoke) may be used.

For biological control of spider mites, there are a number of beneficial mites which kill the red spider mites. The spider mite predators such as *Phytoseiulus persimilis, Neoseiulus californicus and Mesoseiulus longipes* are also very effective and are available commercially. Use of mixtures of these predator mites are recommended to obtain effective control.

Leaf Miner (*Liriomyza trifolii* and *Phytomyza syngenesiae*)

The larvae of this insect make tunnels under the epidermis and disfigure the leaf. The tunnels made by these flies, especially near the margins, are irregular in shape.

Treating the rooted chrysanthemum cuttings with methyl bromide fumigation alongwith cold storage at 1-2°C has been found to be effective against *L. trifolil* eggs and larvae. High dose of nitrogen attracts more leafminers. To control the leaf miners, treat the soil with aldicarb (Temik 10 G) @ 10 kg per hectare. Application of oxamyl (Oxamyl 10 G) was also found to be effective against leaf miner. Use of *Diglyphus begini* and *Dacnusa wasp* are found to be useful to control the leaf miner flies.

White Fly (Greenhouse whitefly-*Trialeurodes vaporariorum*, tobacco whitefly-*Bemisia tabaci*)

Recently the importance of white fly as destructive insect has increased considerably in chrysanthemum cultivation. White fly infestation weakens the plant and in severe cases, even causes death of the plant. Moreover, the white flies are also virus vectors. Yellow sticky traps placed within the greenhouse (atleast one trap per 100 square metre) can help to detect the adult white flies.

To minimize the white fly incidence, weeds growing near the chrysanthemum crops should be removed at the earliest opportunity. Application of endosulfan (Thiodan) @ 2 ml per litre of water or aldicarb (Temik 10 G) @ 10 kg per hectare can control the white flies. Application of imidacloprid (Marathon 1G) for 8-10 weeks has been found to be effective against white flies when used as soil drench or granules. Use of acephate (Orthene), oxamyl (Oxamyl 10 G) and azadirachtin (Azatin) are also reported to be effective against white flies. A number of biocontrol agents such as parasitic wasp *Encarsia formosa*, predatory coccinelid beetle *Delphastus pusillus* and entomopathogenic fungi *Paecilomyces fumosoroseus* have been reported to be effective against white flies. Bakshi et al (2003) have reported the presence of greenhouse white fly (*Trialeurodes vaporariorum*) on chrysanthemum crop in Himachal Pradesh, India.

Tobacco Caterpillar (*Spodoptera litura*)

The full grown catèrpillar is 3.5 cm (approximately) in length and is greenish brown in colour with dark markings. A black ring encircling the body is present at both ends. The larvae feed voraciously and gregariously on the foliage of chrysanthemum plants. These larvae remain hidden in the soil during the day time.

Foliar application of endosulfan (Thiodan) @ 2 ml per litre of water or monocrotophos (Nuvacron) @ 1 ml or carbaryl (Sevin) @ 2 g per litre of water is effective to control these caterpillars. For biological control of these caterpillars, foliar sprays of *Bacillus thuringiensis,* when the caterpillars are small in size, is effective.

Mealybugs (*Phenacoccus* species)

Different types of mealybugs such as Citrus, Greenhouse and Mexican infest chrysanthemums. The Mexican mealybug (*Phenacoccus gossypii*) is primarily a pest of greenhouse chrysanthemums. *P. solani* and *P. madeireusis* also infest the chrysanthemum plants. Severe attack of mealybugs weakens the plant and can kill the plants either directly or indirectly by making the plants susceptible to other ailments. A number of pesticides such as azadirachtin (Azatin), chlorpyrifos (Dursban), bendiocarb (Dycarb), diazinon (Knox out) and acephate (Orthene) are found to be effective when applied at an interval of 2-3 weeks for a number of times. For biological control of this pest, parasites such as *Lindorus lophanthae* and *Metaphyeus helvolus* have been reported to be effective against mealybugs with soft scales.

Chrysanthemum plants are also affected by a number of other insects such as beetle, caterpillars and white grubs.

Nematodes (*Aphelenchoides ritzemabosi, Meloidogyne incognita, M. hapla*)

In recent years, A. *ritzemabosi* has been reported to be very injurious to chrysanthemums grown in the outdoors. The first symptom of infection is yellowish brown spotting of the leaves. The spots enlarge and ultimately the leaves die. The symptoms are very similar to the leaf spot disease caused by the fungus *Septoria*. It is important to note that the spots caused by the nematode are brownish and not black. Severe infection causes death of the plants. *M. incognita* and *M. hapla* also infest the roots of chrysanthemum plants.

A number of precautions should be taken to control the nematode infection. The cuttings should be taken from the shoot tips of the top portions of long and vigorous growing shoots of healthy plants. A crop of marigold should be grown before the cultivation of chrysanthemum. It is better to avoid chrysanthemum cultivation in the same plot year after year. The soil should be thoroughly sterilized. Application of oxamyl (Oxamyl 10 G) either as a soil drench or as granules incorporated with the soil or as a foliar spray has been found to be effective against the nematodes.

Diseases

Chrysanthemum plants are much susceptible to different types of fungal, bacterial and viral diseases. Among the fungal diseases, powdery mildew, damping off, rust, grey mould, *sclerotinia* rot and *septoria* leaf spot cause damages to varying degrees. A number of bacteria and virus, viroids, phytoplasmas also cause extensive damage to the chrysanthemum plants.

Fungal Diseases

Powdery Mildew (*Erysiphe cichoracearum, Oidium chrysanthemi*)

The disease causes white powdery fungal growth on the upper surface of the leaves. In case of severe infection, both the surfaces of leaves and also the stems are covered with white powdery coating. Infection becomes severe in humid atmosphere.

To control this fungal disease, foliar applications of Dinocap (Karathane) or benomyl (Benlate) @ 0.1% may be effective at fortnightly intervals. A number of other fungicides have been found to be very effective against powdery mildew. These are thiophanate-m plus iprodione (Benefit), copper pentahydrate (Phyton-27), fenarimol (Rubigan) and sulfur (Thiolux).

Wani and Ashraf (2003) observed the infection of powdery mildew disease on chrysanthemum plants in Kashmir Valley, India. The disease was found to be severe during May-August and mild to moderate during September-December.

White Rust (*Puccinia horiana*)

This is a very common fungal disease of chrysanthemums. The telia bearing teliospores are mainly on the under surfaces of the leaves but they may also be observed on the upper surfaces of the leaves and stems. Initially the telia are pink in colour, but on maturity, they become white. Hence, the disease is known as white rust. This fungal pathogen forms small blisters on the under surfaces of the leaves. These blisters erupt to release light brown spores. If the infection is severe, all infected leaves may be killed. Warm days and cool nights (10-24°C) favour the spread of rust disease. Cool temperature (10-24°C) favours the spore bearing lesion production.

Control of rust fungi is not easy. In the early stages of rust development, removal of infected leaves can control the disease to a certain extent. During watering, care should be taken that the leaves are not wet, particularly during the evening, because this will encourage the fungal growth. Chemical fungicide sprays can be effective against rust. There are some fungicides such as mancozeb (Dithane) and chlorothalonil (Daconil 2787), which can give protection to the plants against the rust infection. Another group of fungicides is available in the market, which is systemic in action that includes the fungicides such as oxycarboxin (Plantvax), triadimefon (Bayleton) and propiconazole (Banner). As soon as the rust symptoms are visible, the infected plants should be sprayed with oxycarboxin (Plantvax) @ 1 g per litre of water or mancozeb (Dithane M-45) @ 2 g per litre as a foliar spray.

The symptom of chrysanthemum white rust is discoloured spots on the leaf surface with pustules containing the spores of the causal agent *Puccinia horiana* on the under surface of the leaves. Preventive measures include avoiding hot damp conditions under cover and not watering in the evening. For rust prone areas, resistant chrysanthemum varieties should be used. The fungicides such as triforine, bitertanol, penconazole, mancozeb and chlorothalonil are found to be effective. Myclobutanil has been used successfully to eradicate infection but should not be used routinely as there is a risk of building up resistance. Propiconazole may cause injury in some chrysanthemum varieties (Riccioni 2004).

Verticillium Wilt (*Verticillium albo-atrum*)

Verticillium wilt and another wilt of chrysanthemum caused by *Fusarium oxysporum* var. *tracheiphilum* are very destructive fungal diseases of chrysanthemum. The initial symptom is conspicuous yellowing and browning of the leaves and this discolouration progresses upwards. Infected plants become stunted and many a times fail to produce flowers. Wilt fungi are soil borne.They enter the plant through the roots and clog the vessels of the stem and disrupt the water supply. Greenhouse chrysanthemums are usually susceptible to this disease but sometimes the outdoor-grown, hardy chrysanthemums are also infected.

To prevent and control this fungal disease, the soil should be sterilized by using methyl bromide or chlorpicrin. Disease-free healthy cuttings should be used and preferably be planted on raised beds or benches.

Septoria Leaf Spot (*Septoria chrysanthemi* and *S. chrysanthemella*)

Circular, grey brown spots are formed on the older leaves, which become dark brown and then black in due course of time. In severe cases, it may result in withering of leaves. The lower leaves are infected first. With the help of a hand lens, white masses of spores may be seen on the minute specks, which are numerous in the diseased area. The spores are long and slender.

A number of other fungi also cause leaf spots on chrysanthemum plants. Most prevalent of these are *Alternaria chrysanthemi, Cercospora chrysanthemi, Phyllosticta chrysanthemi* and *Cylindrosporium chrysanthemi.* The latter produces dark brown spots with yellowish margins. These spots gradually increase in size and affect the whole leaf, which dies and hangs down on the stem like the *Septoria* infected leaves.

To control these diseases, the diseased plants should be destroyed as early as possible. Weekly applications of mancozeb (Dithane M-45), zineb (Dithane Z-78) or orthocide (Captan) can provide good protection to the plants. A number of other fungicides such as thiophanate-m plus iprodione (Benefit), iprodione (Chipco 26019), vinclozolin (Curalan) and chlorothalonil (Daconil 2787) are found to be effective against these fungal pathogens.

The efficacy of Indofil M-45 (mancozeb 75 WP at 0.25%), Dithane Z-78 (zineb 75 WP at 0.25%), Captaf (captan 50 WP at 0.25%), Bavistin (carbendazim 50 WP at 0.10%) and Companion (carbendazim 12+ mancozeb 63 WP at 0.10%) against *Septoria* leaf spot (*Septoria* species) has been evaluated under field conditions in West Bengal, India. The results revealed that Companion gives the highest disease control (77.3%) followed by Bavistin (69.8%) and Captaf (65.8%). Dithane Z-78 is the least effective among the fungicides (Jash et al 2003).

Ray Blight (*Didymella chrysanthemi, Mycosphaerella ligulicola*)

In this case, the ray florets of the chrysanthemum flowers become deformed on one side and the whole flower becomes one sided due to the infection. Early infection by these fungi may cause blasting of the flower buds. The infection of inner florets renders them dark brown in colour and initiates rotting. To control this fungal disease, the infected plants should be removed. Foliar sprays of benomyl (Benlate) and mancozeb (Dithane M-45) @ 2 g per litre of water are very effective against this disease.

Gray Mold (*Botrytis cinerea*)

Infection of this disease leads to brown water soaked spots on the petals of the buds and flowers. Infected plants get covered with a greyish brown powdery mass of spores. This disease spreads from plant to plant at the time of watering and through air currents. In severe cases of infection, the flower buds and flowers may rot.

To control the disease, the ventilation system within the greenhouse should be improved so that there is free flow of air. Watering should be done in the morning hours so that the leaves of the plants dry before afternoon. Foliar sprays of benomyl (Benlate) or orthocide (Captan) or mancozeb (Dithane M-45) @ 2 g per litre of water are effective to control this disease.

Root Rot and Damping Off (*Pythium* species, *Phytophthora* sp.)

Root rot of chrysanthemum is caused by *Pythium* and *Phytophthora* sp. under moist warm conditions. Rotting of stem cuttings due to infection in rooting bed is a common sight. The fungi also enter the established plants through wounds caused by pinching. Sudden wilting of infected parts (root, stem and leaf) is the characteristic feature of this fungal disease.

To control this fungal disease, the rooting medium should be properly sterilized before planting the cuttings. Any infected cutting should be removed immediately. After planting, soil drenching with orthocide (Captan) or ethidiazole plus thiophanate-m (Banrot) or metalaxyl (Subdue) is very effective as a control measure.

Sclerotinia Rot (*Sclerotinia sclerotiorum*)

Infection by this fungus produces light brown lesions on the stem. These lesions rapidly girdle the stem and cause wilting of the plants. A dense white fluffy fungal growth appears on the infected part of the plants during wet weather.

To control the disease, the soil should be sterilized by using methyl bromide or chlorpicrin or formaldehyde. Foliar sprays of carbendazim (Bavistin) @ 1.5 g per litre of water or benomyl (Benlate) @ 2.0 g per litre are effective to control this fungal disease.

Beneficial Fungi

A number of fungi have shown beneficial effects on plant growth and flowering of chrysanthemum plants. Arbuscular mycorrizal fungi (VAM) showed high mycorrhizal association with chrysanthemum (Gupta and Routaray 2004).

Bacterial Diseases

Chrysanthemums are also prone to a number of bacterial diseases. High temperature (27-32°C) and humidity favours disease development and spread, and results in wilting of plants. With the spread of the disease, the stem tip turns brown, the stem becomes hollow and brittle, and ultimately the plant collapses. In rooting beds, the infected cuttings either die or remain stunted. To control these bacterial diseases, use of disease free cuttings, sterilization of soil and spraying with antibiotics such as streptocycline or terramycin (0.01%) are recommended.

Bacterial Blight (*Erwinia chrysanthemi, E. carotovora*)

The salient feature of this bacterial disease is rotting of upper part of the stem which results in wilt and collapse of the distal portion. Cuttings infected with this bacterium may show a brown to black decay at the base. Sometimes the only symptom of the disease is a marginal leaf scorch.

To control this disease, the soil should be sterilized before planting the cuttings. A 4-hours dip in antibiotic solutions of chloromycetin, streptomycin and terramycin has been reported to be an effective control.

Leaf Gall or Fasciation (*Corynebacterium fascians*)

Due to this bacterial infection, the stems become short and thick near the crown, and results in aborted and deformed leaves. Due to infection, the plant becomes dwarf. The plants are infected at the early stage of the growth but the disease symptoms are expressed when the plants mature.

The soil should be disinfected by using methyl bromide or chlorpicrin or formaldehyde to prevent this disease.

Bacterial Leaf Spot (*Pseudomonus cichorii*)

The spots produced by this disease are dark brown to black in colour, slightly sunken and have concentric areas. The infected plants should be treated with streptomycin (Agrimycin) or a copper compound such as cupric hydroxide (Kocide) or copper pentahydrate (Phyton-27).

Viral and Viroid Diseases

Chrysanthemum plants are also suceptible to a large number of viral diseases. Prominent among these are tomato spotted wilt virus (TSWV), tomato aspermy virus (TAV), chrysanthemum virus B (CVB), turnip mosaic virus (TuMV) and cucumber mosaic cucumovirus (CMV).

Tomato Spotted Wilt Virus (TSWV)

The infection produces ring and line patterns, pale areas, mottling, and necrotic spots. TSWV is localized in symptomless stock plants and in the cuttings taken from them. Its detection frequency is relatively high in the lower leaves and stems but low in terminal buds. The transmission efficiency from infected chrysanthemum stock plants to cuttings is fairly high (Matsuura et al 2004).

Turnip Mosaic Virus (TuMv)

During the summer of 2003, ornamental crops grown in seven commercial greenhouses in Markazi, Iran showed signs of mottling on *Chrysanthemum* sp. The isolated pathogen was identified as turnip mosaic virus (Farzadfar et al 2005).

Tomato Aspermy Virus (TAV)

This virus causes distortion and abnormal colouration of chrysanthemum flowers. In nature, this virus is transmitted by a number of aphids such as *Myzus persicae, Macrosiphoniella sanborni and Acrythosiphon solani.*

Chrysanthemum Virus B (CVB)

An indirect enzyme-linked immunosorbent assay, used for indexing chrysanthemum viruses, indicated that CVB is among those which are widely prevalent. Disease prevention by using only disease-free indexed mother plants is the best option. Control of vector population by regular application of malathion @ 2 ml per litre water or endosulfan @ 2 ml per litre is helpful in restricting the disease spread.

Phytoplasma

Aster Yellows

This disease caused by a phytoplasma like organism, is characterized by the production of green coloured flowers instead of the normal colour for the variety. Sometimes the upper branches of an infected flowering stem are thin, yellowish and more upright than usual. The diseased plants die within a few months of infection and should be removed. Foliar sprays of malathion (Malathion) or endosulfan (Thiodan) should be carried out to control the vector population. Aster yellows phytoplasma infected chrysanthemums (*D. morifolium*), showing stunt, rosette and excessive branching, when treated with a foliar spray of 400 mg per litre oxytetracycline did not express any phytoplasma infection symptoms for more than 10 months. Chrysanthemums dipped in 100 mg per litre oxytetracycline solution combined with a foliar spray of 400 mg per litre oxytetracycline for 4 weeks also showed the same results.

HARVESTING AND YIELD

The correct stage of harvesting in chrysanthemum depends on the cultivar, distance from the market, and facilities available to the grower. The spray types are harvested when the flowers are half opened. This also facilitates in packaging. Anemone cultivars are harvested before the central disc in the top most flower is fully developed. The decorative types should be harvested when the petals in the centre of the top most flowers are almost fully developed. In standards, harvesting is generally done when outer ray florets cease to elongate.

Cutting of flowers should be done about 10 cm above the soil and one third of stems should be stripped off the leaves. The cut flowers should be placed immediately in water containing a biocide to prevent the growth of microorganisms. For the purpose of garlands and other floral crafts, flowers should be harvested without stems. If the flowers are harvested at a slightly immature stage, use of bud opening solution become very useful. An ideal bud opening solution is 150 ppm 8-hydroxyquinoline citrate (8-HQC) and 2% sucrose. The performance of another bud opening solution comprising 3% sucrose, 25 ppm silver nitrate and 75 ppm citric acid was also found to be effective.

About 0.60-0.80 lakh good quality sprays are usually harvested from an acre of land whereas 1.0-1.25 lakh standared flowers are produced from the same area.

Grading

Cut flowers are grouped in several grades depending on the stem length and strength, colour and diameter of flowers. Stem length should not be less than 60 cm as flowers with shorter stems fetch lesser price.

Packaging

Spray chrysanthemums having 10, 15 or 20 stems are placed in sleeves according to the grades. Six sleeves, three at each end, are generally packed in each box measuring 80 x 50 x 23 cm. Most of the standard chrysanthemums are placed in sleeves and packed in display boxes measuring 90 x 45 x 15 cm (approximately). In all display packs, it is better to hold stems in place with a metal tie strip.

Storage of Flowers and Cuttings

Chrysanthemum cut flowers can be wrapped in plastics and stored in dry places for 6 to 8 weeks at a temperature of –0.5°C. A pre-storage dip in contact insecticide for half an hour is effective in controlling pests during long term storage. Refrigerated vans should have a temperature between 2-4°C while carrying the flowers. Beneficial effects of light during storage can be achieved by raising the light intensity to 4000 lux. Storage in dark deteriorates the quality of the bloom.

Post-harvest Technologies

The presence of old stubs on flower stems affect the uptake of water and thus should be removed. Immediately after harvesting, the cut stems should be placed in a bucket containing preservative solution to extend the vase life of the cut flowers. Dipping of cut stems in 1000 ppm silver nitrate solution for 10 minutes has been found to be very effective to prolong the vase life. Cut flower stems treated with 4% sucrose solution also is found to be beneficial.

Post-harvest changes affect the commercial viability of the cut flowers. One of these changes is the fresh weight of the flower as influenced by its water balance. Placing the cut flowers of chrysanthemum in preservative solution of cobalt nitrate is helpful in retaining fresh weight and water balannce (Patel and Mankad 2002).

CULTIVARS FOR CUT FLOWER PRODUCTION

Spray Cultivars

White - Reagan White, Majoor Bosshardt, Cassa, Daymark, Giraffe, Improved Penny Lane, White Mundial, White Moneymaker, Remark, Spinmark, Spider White, Stallion, Starmark, Super White, White Illini Springtime, White Marble, Bonnie Jean, Elegans, Cloudbank, Arctic, Polaris, White Stafour, Inga

Yellow - Reagan Yellow, Reagan Sunny, Spider Yellow, Regoltime, Lier Way, Your Way, Free Way, Butter Cup, Greta

Verhagen, Sunny Casa, Golden Casa, Toon Hermans, Town Talk, Feeling, Yellow Bosshardt, Yellow Mundial, Peach Mundial, Apricot Moneymaker, Yellow Daymark, Apricot Impala, Golden Westland, Sulphur Westland, Yellow Westland, Hippo, Funray, Main Lane, Golden Crystal, Golden Hurricane, Super Yellow, Yellow Divinity, Sunbeam

Bronze - Apricot Marble, Apricot Dinner, Dramatic, Bronze Mundial, Golden Bronze Moneymaker, Golden Bronze Impala, Golden Bronze Byoux, Bronze Nero, Bronze Roserdo, Belreef, Galaxy, Tuneful, Flame Belair

Salmon - Reagan Salmon, Reagan Currant, Coral Marble, Salmon Mundial, Mundial, Salmon Byoux, Park Lane, Impala

Pink - Reagan Improved, Reagan, Glance Improved, Pink Marble, Belair, Della Street, Bond Street, Pink Mundial, Royal Mundial, Coral Moneymaker, Pall Mall, Bay Frank, Dark Bonus, Regal Westland, Pink Impala, Pink Daymark, Funshine, Improved Funshine, Franky Lane

Red - Reagan Red, Red Fandango, Red Galaxy, Red Nero, Red Tuneful, Crackerjack, Carnaby Street, Tiger Rag

Purple, Lilac - Fandango, Flamenco, Lilac Byoux

Orange - Reagan Orange, Tiger, Reagan Splendid, Reagan Dark Splendid, Delilah, Harlequin

Large Flowering Cultivars

White - Snow Ball, General Petain, Casa Grande, William Turner, Innocence, May Shoesmith, Valiant, Premier, Giant Indianapolis White, Improved Mefo, Snow Don, Ajina White, Beauty, White Harmony

Yellow - Kikubiori, Mountaineer, Chandrama, Super Giant, J.S. Lloyd, Sonar Bangla, Triumphant, Evening Star, Bright Golden Anne, Bright Yellow May Shoesmith, Yellow Fred Shoesmith, Rivalry, Melody Lane, Duke of Kent, Coronation Gold, Lady Frank Clarke, Betty Baarnes, Yellow Snowdon

Pink - Cassandra, Deep Champagne, Pink Champagne, Promenade, Regal Anne, Dream Castle, Cover Girl, Pink Turner, Pink Cloud, Fish Tail, Pink Gin

Purple - Ajina Purple, Royal Purple, Purple Anne

Bronze - Bronze Princess Anne, Resilient, Gay Anne

Red - Crimson Anne, Red Anne, Red Resilient, Woking Scarlet, Guesman Red, Alfred Wilson, Alfred Simpson, Mrs. W.A. Reid

Green - Green Goddess, Woolman's Century, Madam E Rogers

REFERENCES

1. TK Bose, LP Yadev, P Pal, VA Parthasarathy and P Das (eds) 2002 Chrysanthemum, in *Commercial Flowers* (Volume 1), Calcutta, India: Naya Prakash, 465-602 pages
2. Carvalho SMP, Heuvelink E, and van-Kooten O 2002 Effect of light intensity, plant density, and flower bud removal on the flower size and number in cut chrysanthemum. *Acta Hortic.* **593** 33-38
3. Cockshull KE 1985 *Chrysanthemum morifolium,* in *CRC Handbook of Flowering* (Volume II) AA Halevy (ed), Florida, USA: CRC Press Inc., 238-257 pages
4. Cumming RW 1964 *The Chrysanthemum Book*, Princeton, New Jersey, USA: D. van Nostrand Co. Inc.
5. Farzadfar S, Ohshima K, Pourrahim R, Golnaraghi AR, Jalali S and Ahoonmanesh A 2005 Occurrence of Turnip mosaic virus on ornamental crop in Iran, *Pl. Path.* **54** 261
6. Gupta N and Routaray S 2004 Arbuscular mycorrhizal association of weeds found with different plantation crops and nursery plants, *Mycorrhiza News* **16** 14-20
7. Jash S, Bandyopadhyay S, Dutta S and Laha SK 2003 Comparative efficacy of fungicides against Septoria leaf spot of chrysanthemum under terai agro-ecological region of West Bengal, *J. Mycopathological Res.* **41** 109-110
8. Korner O and Challa H 2004 Temperature integration and process based humidity control in chrysanthemum, *Computers & Electronics in Agril.* **43** 1-21
9. Machin BJ and Scopes N 1978 *Chrysanthemum-year round growing*, Dorset, U.K: Blandford Press
10. Matsuura S, Ishikura S, Shigemoto N, Kajihara S and Hagiwara K 2004 Localization of tomato spotted wilt virus in chrysanthemum stock plants and efficiency of viral transmission from infected stock plants to cuttings, *J. Phytopathology* **152** 219-223
11. Mukherjee D, Raja Ram and Dhyani D 1994 Studies on growth and flowering of different varieties of chrysanthemums under greenhouse and open field conditions, 21-27. XXIV International Horticultural Congress, August, Kyoto, Japan (Abstract No- 210).
12. Riccioni L 2004 Chrysanthemum white rust, quarantine pathology, *Colture-Protette* **33** 79-81

13. Sun ZF, Li M, Chen L, Li SR and Li CS 1999 Studies on the forcing culture of cut summer flowering chrysanthemum in solar greenhouse, *Acta-Hortic. Sinica* **26** 391-396

14. Wani AH and Ashraf M 2003 Incidence and severity of powdery mildews of some plants of Asteraceae in some famous gardens of Kashmir valley, *Bioved* **14** 15-19.

MARIGOLD

S.P.S. Raghava, Kanwar Pal Singh and V. S. Raju Dantuluri

INTRODUCTION

Marigold is one of the most important traditional flower crops grown commercially in different parts of India. It has gained popularity because of its adaptability to various soils and climatic conditions, longer blooming period, and beautiful flowers, which have good shelf life. Marigold is grown for loose flower production. The flowers are used extensively for making garlands, for beautification of mandaps and decoration of cars in marriages and religious offerings. Marigolds are grown in beds as well as in pots. Some varieties of African marigold, which have straight stalk and globular flowers are used as cut flower for interior decoration also. The cultivation of marigold is quite remunerative on account of its various commercial uses. The inventions carried out in marigold have resulted in giving a good fillip to its utility in different spheres. The acreage in India under commercial cultivation of marigolds has gone up dramatically.

INDUSTRIAL AND OTHER USES

Recently, marigold has emerged as a source of carotenoid pigments, which are widely used as dietary supplements in poultry industry to enhance chicken skin colour and egg yolk pigmentation. The major carotenoid in marigold is xanthophyll. The high solubility of lutein fatty acids in vegetable oil is comparable to the FDA approved synthetic carotenoids. This is an important factor in the commercial application of xanthophyll esters for colouring food. It is reported that purified extract of marigold petals, mainly containing xanthophyll dipalmitate is marketed as an opthalmological agent under the trade name ADAPTIONAL.

The essential oil present in different species of *Tagetes* is used in perfume industry. *Tagetes signata* has the highest oil content (4.25% of dry matter) of very good quality, which can be used in the perfume industry. *Tagetes minuta* is used commercially for the extraction of the essential oil. France, Kenya, Australia and Brazil are the main essential oil producing countries in the world. This species is found in natural habitat in the States of Himachal Pradesh, Uttarakhand and Jammu and Kashmir states in India. It has become a cash crop in the hills of Himachal Pradesh. The oil produced at higher altitudes is rich in ketonic constituents, while the one that is cultivated in the plains of Punjab is of low olfactory value. On an average, 20-25 tonnes of biomass harvested per hectare yields 50-60 kg oil. The yield of oil varies from 0.20-0.35%. In the recent past, *Tagetes minuta* oil has been identified as a valuable and precious essential oil for use in high grade perfumes and cosmetics. The floral oil and absolutes are used for compounding high grade perfumes. Ketones found in high concentration in the oil are the base material for synthesizing valuable aroma chemicals. The oil has juvenile hormonal and insect repellant activities against flies, ants and mosquitoes.

Marigold is also used in the traditional Mexican medicines as well. Both leaves and flowers are equally important from medicinal point of view. Leaf paste is used as topical application for boils and carbuncles. Leaf extract is a good remedy for earache. The oil has bronchodilatory, tranquilizing and anti-inflammatory properties. Marigold flower extract is considered as blood purifier, a cure for bleeding piles and a good remedy for ulcers.

Marigold produces aromatic, sulfur containing, secondary compounds, mainly in roots, known as thiophenes, which have been found to be effective in keeping the nematode population under control. Marigold is planted as an antagonistic crop in vegetable crops to control nematodes.

ORIGIN AND HISTORY

Marigold is native to South and Central America, especially Mexico. From Mexico, it spread to different parts of the world in 16th Century. The generic name *Tagetes* was derived from the Latin word t 'Tagtes' the name of an Entruscan divine, the God of beauty. Though native to American continent, marigold was discovered by the Spanish traveller Cortz who introduced it to Europe. Devotees in Spain used to place these flowers at the altar of Virgin Mary and from that day they came to be known as Mary's gold or marigold. The two important species got the names African marigold and French marigold, although the original species from which they have been bred were introduced from America.

African marigold was first introduced into Spain early in the 16th century and became popular in Europe in 1585 under the name Flors Africonus by Emperor Charles V, who launched an expedition to free Tunis from the Moors where it was naturalized on the Algerian coast. It continued to be called so until 18th Century, although it was correctly figured as Plantas *Tagetes indica* in Fuch`s Herbal of 1542.

Another well-known type of marigold, the French marigold, followed the route similar to that of African marigold. Parkinson mentioned that it was called *Tanacetum peruvianum* as it grew in Peru. James Justice believed that it is a Chinese plant and it was first sent to the Royal Garden at Paris from where it was distributed under the name, French marigold.

BOTANY

Marigold belongs to the family Asteraceae (Compositae). In fact, each flower in marigold is composed of a mass of small flowers called florets. What we generally understand to be a flower is a complete inflorescence called head or capitulum. Two forms of florets exist in most of the flowers of the family Asteraceae. Those on the outside of the disc are called ray florets and the others which comprise the disc are called disc florets. Ray florets are female and zygomorphic. The five florets are fused to form a ligule or a flat petal. The disc florets are tubular, actinomorphic and bisexual. The double forms are generally male sterile as the disc florets are converted into showy petals. Double forms are ornamentally preferred and have good economic value

The basic chromosome number of Marigold is 12. *Tagetes erecta, T. tenuifolia* are diploid having a chromosome number of 24. *Tagetes patula* and *Tagetes minuta* are tetraploids having a chromosome number of 48. *Tagetes patula* is an allotetraploid between *Tagetes erecta* and *Tagetes tenuifolia*. Unlike other plants, where the size of the plants and flowers increase with increase in ploidy level, in marigold it decreases exceptionally.

SPECIES

There are about 33 species of the genus *Tagetes*. The characters of important species are given below:

***Tagetes erecta* (African marigold):** The plant is a hardy annual. It grows to a height of 70-100 cm. The plant type is erect. The leaves are pinnately divided, and the leaflets are lanceolate and serrated. Flower type varies from single to double. The flowers are globular. The florets are either 2-lipped or quilled. Flower colour varies from sulfur yellow to orpiment orange.

***Tagetes patula* (French marigold):** The plant is a hardy annual. It grows to a height of 30-45 cm. The stem is reddish and the foliage is dark green. Leaves are pinnately divided, and the leaflets are linear, lanceolate and serrated. Flowers are yellow to brick red in colour. In most of the cases, brick red coloured blotches are seen on yellow background.

***Tagetes tenuifolia* (Signet marigold):** This species is an annual with branching habit. It is widely distributed from North America to

Columbia. Leaves are pinnately divided into 12 oblong, linear, sharply serrated segments. Flowers are yellow in colour and numerous. Flower heads are small, 2.5 cm across. Flowers have five rays, which are roundish and obovate.

***Tagetes lucida* (Sweet scented marigold):** The plants of this species are perennial. Leaves are sessile, small and lanceolate. Flowers are usually 2-3 rayed, produced in dense terminal corymbs.

***Tagetes minuta*:** This species is an annual that grows 0.9-1.0 m tall. Leaves are pinnate. It produces small white flowers in dense terminal clusters. Oil is extracted from this species commercially.

VARIETIES

Many varieties and strains of marigold are available that vary in plant height, growth habit, flower shape, colour and size. Mostly local varieties are being cultivated by the farmers.

African Marigold

Pusa Narangi Gainda: It is a cross between 'Cracker Jack' and 'Golden Jubilee'. It flowers in 125-136 days after sowing. Its plants are 73.30 cm tall, vigorous and uniform with dark green foliage. Flowers are orange in colour, carnation type, double, compact, 7.8 cm across with disc florets. It is most suitable for garland making, religious offerings and carotenoid extraction.

Pusa Basanti Gainda: It is a cross between 'Golden yellow' and 'Sun Giant'. It is a late maturing variety, which takes 135-145 days. The flowering duration is long (40-45 days). The plant is 60 cm tall, vigorous and uniform with dark green foliage. Flowers are sulfur yellow in colour, carnation type, double, compact, 6.9 cm across with disc florets but invisible. The number of whorls present in the capitulum is 7-10. It is ideal for loose flower production and also for growing in pots and beds in the garden.

Inca Hybrid: These plants are 30-40 cm tall and produce 5 inches balls that radiate vibrant yellow colour, and they completely blanket the vigorous plants all the season. The flowers are fully double and solidly packed with waxy, tightly-furled petals.

Crackerjack: Flowers are gold, yellow and orange in colour, 10 cm across and ruffled. The plants grow to a height of 1m.

Giant Double African Orange: These plants are 90-100 cm tall, with flowers that are double, orange coloured, compact, and carnation type.

Giant Double African Yellow: These plants are 90-100 cm tall. Flowers are double, yellow in colour, compact and carnation type.

Chrysanthemum Charm: These plants are 70-80 cm tall. Flowers are double, chrysanthemum type and yellow in colour.

Spun Gold: These plants are 80-90 cm tall. Flowers are double and golden in colour.

Cupid Yellow: These plants are 60-70 cm tall. Flowers are double, chrysanthemum type and yellow in colour.

Ladoo Orange: These plants are 90-100 cm tall. Flowers are double, medium sized and orange coloured.

Ladoo Yellow: These plants are 90-100 cm tall. Flowers are double, medium sized and yellow in colour.

French Marigold

Pusa Arpita: The variety abundantly produces flowers of yellow colour and medium size. It is suitable for loose flower production. Flowering is during December-January.

Red Brocade: These plants are 20-30 cm tall. Flowers are semi-double and red in colour. It is suitable for pots and garden display.

Rusty Red: These plants are 20-30 cm tall. Flowers are semi-double and maroon in colour. It is suitable for pots and garden display.

Harmony: These plants are 30-40 cm tall. Flowers are semi-double and petals have deep orange centre with an edge of crimson.

Star of India: These plants are 30-40 cm tall. Flowers are deep crimson in colour, striped and marked with golden yellow.

Naughty Marietta: Plants are 30-40 cm tall. Flowers are single and petals are blotched with maroon colour at the centre.

Lemon Drop: These plants are 30-40 cm tall. Petals are double and lemon in colour.

Interspecific Hybrids

The interspecific hybrids produced by crossing African marigold and French marigold are intermediate in characters. They exhibit the floriferousness of African marigold and the branching character of French marigold. The interspecific hybrids are triploids, as a result they are sterile, and no seed formation is observed. These hybrids are favoured by private seed companies as they prevent illegitimate seed production. Some of the popular interspecific hybrids are Show Boat, Nuggets and Red Seven Star.

SOIL AND CLIMATE

Marigold is adaptable to different types of soil conditions and thus, can be cultivated successfully in a wide variety of soils. A deep fertile, friable soil having good water holding capacity and aeration is suitable for Marigold. Sandy loam soil with a pH of 7.0-7.5 having good drainage is ideal for its cultivation.

Marigold can be cultivated throughout the year. During the hotter months (under long days) in North Indian conditions, flower production is affected adversely and flower size is also reduced. Marigold needs plenty of sunshine, so it should be grown in open sunny conditions. In winter, the plants and seedlings are damaged by severe frost. The environmental conditions markedly influence the growth and flowering of marigold. Mild temperature during growing period greatly improve the flowering. The flowering of marigold can be taken throughout the year by taking crops in succession in order to ensure their regular supply to the market.

PROPAGATION

Marigold is commercially propagated by seeds. It is also propagated by cuttings by some farmers to obtain uniformity in growth and flowering. Male sterile lines are also propagated and maintained by cuttings. Plants raised from seeds are vigorous, healthy and they establish well.

Seed Propagation

Seeds of many cultivars are easily available and germinate quickly. Marigold seeds germinate well at temperatures ranging from 18-30°C. Nursery beds are prepared after thoroughly mixing well decomposed FYM. Raised nursery beds of convenient size should be prepared. About 1 kg of seed is required for one hectare. Before sowing, the seeds should be treated with Captaf or Bavistin. The nursery bed may also be drenched with 0.1% Bavistin. Seeds should be finely covered with FYM or leaf mould. Depth of sowing should not exceed 2-3 cm. Watering should be done with the help of a fine rose can. Formation of hard pan should be prevented for good emergence of seedlings. The sowing should not be too dense to prevent damping of the seedlings.

Propagation through Cuttings

Marigold is propagated through cuttings also to produce true to type plants and to maintain male sterile lines. Cuttings measuring 8-10 cm in length, treated with root promoting substances such as IBA and NAA are planted in coarse sand for rooting. After 2-3 weeks, sufficient roots are formed on the cuttings and they can be transplanted in the field. This method of propagation is most suited for completely double flowered plant types, which do not set seeds.

Seed Sowing Time

In plains, marigold can be cultivated throughout the year. In North India, the ideal time for winter crop sowing is second week of September and for summer crop sowings, it is second week of February. Sowing during second week of June is optimal for rainy season crop. However, marigold crop can be raised two times in hills. Sowing of seeds for first crop can be done in March and the seeds for second crop can be sown in May.

PLANTING

The seedlings should be transplanted in the field one month after seed sowing. A spacing of 45 x 45 cm for African marigold and 30 x 30 cm for French marigold has been found to be optimum for commercial cultivation. The seedlings at the time of transplanting should be disease-free and should have 3-4 true leaves. After transplanting, a light irrigation should be given.

Nutrition

Fresh and dry weight of plants increases for 70-75 days until flowering and need for nitrogen lasts till this stage. Potassium is required till seed formation (90-100 days). Well decomposed FYM @ 30 tonnes should be incorporated per hectare in the soil at the time of soil preparation. In addition to FYM, it is advisable to apply 120 kg per hectare of nitrogen and 80 kg per hectare each of phosphorus and potash for getting good vegetative growth and flowering. Whole quantity of phosphorus and potash should be applied at the time of land preparation. Nitrogen should be applied in two split doses. The first dose should be applied 20 days after planting and the second dose 40 days after transplanting.

Pinching

In African marigold, because of apical dominance the plants grow tall and lanky with very few branches. Such plants are susceptible to lodging and the flower production is less. Pinching is done to break apical dominance and to encourage axillary branching and thereby, increasing flower production. French marigold does not require pinching. Pinching slightly delays flowering but improves flower production. Pinching 40 days after transplanting has been found to be optimal.

Growth Regulators

The cultivars of African marigold usually grow taller. Growth retarding chemicals have been recommended for reducing the height and improving the appearance of these as potted plant. SADH 2500 ppm reduces plant height and improves foliage colour. When 2-5% solution of CCC is sprayed, there is a marked improvement in growth and flowering.

The treated plants are dwarf in nature. Repeated applications of GA_3 slightly improve height and flowering.

Irrigation

For all stages of vegetative growth and flowering, sufficient amount of moisture is essential. Moisture stress at any of the vegetative and flowering stages influences growth and flowering in a negative way. The frequency and quantum of irrigation is dependent on growing area, soil and season. During summer months in the northern plains, the plants are liable to desiccation due to hot winds. Therefore, the plants should be irrigated according to the prevailing atmospheric conditions. Normally, marigold requires irrigation at an interval of 8-10 days.

Weed Control

Weed control in marigold is labourious and it involves lot of cost. If weeds are not controlled in time, growth, flowering and productivity are drastically affected. Atleast 3-4 manual weedings are required for the entire cropping period. Alachlor @ 2.5-3.0 kg per hectare is effective in controlling the weeds during the entire cropping period.

Harvesting and Packing

The flowers are harvested when they are fully open. The flowers should be plucked either in the morning or in the evening by hand. After harvesting, flowers should be kept in shade under proper ventilation. They should be graded on the basis of their size. The bruised and damaged flowers should be discarded. The packing of flowers should be done in bamboo baskets or gunny bags.

The commercial varieties of African marigold yield about 20-24 tonnes of flowers per hectare whereas varieties of French marigold yield about 8-12 tonnes per hectare.

Pigment Extraction

Commercial extraction of pigments in marigold is done in Cochin and Hyderabad. There are large areas in Karnataka, Andhra Pradesh and Maharashtra where cultivation of marigold is done on a contractual basis for pigment extraction. Marigold petals are the most common concentrated source of xanthophylls. Pigments have been extracted from *Tagetes erecta, Tagetes patula, Tagetes grandiflora* and *Tagetes nana*. The petals of orange colour flowers are best for extraction of carotenoids because of higher content and ease of extraction. Analysis of flower petals of *Tagetes erecta* for total carotenoids at the Indian Agricultural Research Institute, New Delhi revealed a variation from 1.6-4.5 mg per gram of petals on fresh weight basis in different varieties. Dried marigold petals contain approximately 1.6% carotenoids. The major carotenoid fraction in marigold flower petal is xanthophyll. Lutein accounts for 80-90% of the total xanthophyll.

Extraction Procedure

Commercial scale extraction of carotenoid pigments from flower petals is done by using organic solvents such as hexane. Flowers are heaped and covered with black polythene sheet and allowed to ferment for 30-45 days. After fermentation, the flowers are dried, powdered and the powder is mixed with hexane. The pigment dissolves in hexane. Later, the volatile solvent is allowed to evaporate and the total pigment is recovered.

Oil Extraction

Essential oil is present in the leaves, flowers and seed pods only, whereas stem contains little or traces of the oil. It is found that marigold crop should be harvested at full bloom stage around October or early November in Himachal Pradesh hills. On an average 20-25 tonnes per hectare of herb biomass is harvested, which in turn yields 50-60 kg per hectare oil. The essential oil from the crop is preferably obtained by steam distillation. The harvested crop should be allowed to wilt for 3 days in shade and then distilled in hydro-distillation and steam distillation units. The oil yield varies from 0.20-0.35%. The technology of extraction unit is available with Institute of Himalayan Bioresource Technology, Palampur (Himachal Pradesh, India).

Seed Production

Seeds of important varieties are produced by maintaining isolation to obtain pure seeds. The seed produced should have purity, vigour and good germinability. Specific standards for marigold seed production have not been specified so far. Marigold for seed production should be grown in open sunny conditions. The soil should be fertile, well drained with a pH of 7.0-7.5. Marigold is a cross-pollinated crop and pollination is aided by insects like honeybees. For seed production, an isolation distance of 500 metre is maintained between varieties. About 1000 metre isolation distance is maintained for nucleus and breeder seed production.

Harvesting of fully dried flowers is done. At the time of picking, flowers in which seed setting has not taken place should be discarded. The dried flowers are thrashed and winnowed to obtain the seeds. The yield of the cleaned seeds is 100 kg per hectare. The seeds can be stored in muslin cloth bags or paper bags in a well-ventilated place.

INSECT PESTS AND DISEASES

Insect Pests

Red spider mite (*Tetranychus* sp.): Mites infect old plants. The infected plants give dusty and webbed appearance. Mites can be controlled by spraying Kelthane (Dicofol) 0.1%.

Hairy caterpillar (*Diacrisia obliqua*): In the initial stages, they eat away the foliage and at later stages they eat away the ray florets of the inflorescence. These can be controlled by spraying Nuvan (0.2%).

Leaf hopper (*Empoasca fabae*): The hoppers suck sap from the leaves and stem. The infected plants show cupped or rolled leaves and give wilted appearance. Spraying of Metasystox (0.2%) controls the hoppers.

Slugs: They feed on the leaves of marigold. They can be controlled with 0.2% Metaldehyde.

Diseases

Damping off (*Pythium* sp., *Rhizoctonia solani*): This disease affects the seedlings. Before emergence, brown coloured spots appear on radicle and plumule causing pre-emergence death of the seedlings. After emergence, brownish necrotic spots appear on the collar portion of the seedlings. These necrotic spots, in severe cases, form a girdled portion leading to collapse of seedlings. The disease can be controlled by prophylactic drenching of nursery beds with Captaf (0.1%) and Bavistin (0.1%). Seed treatment with Captaf 0.1% can also reduce the incidence of the disease.

Leaf spot and blight (*Alternaria tagetica., Cercospora tagetis, Septoria tageticola, Colletotrichum capsici.*): Various pathogens cause leaf spot and blight in marigold. The symptoms appear as small brownish spots on leaves. In case of *Alternaria* infection, the spots show concentric circles. In severe cases, the infection may spread to inflorescence. In case of *Cercospora,* the spots show bird eye appearance with dark border and light coloured centre. *Septoria* infection is marked by black spots. The blight of marigold is caused by *Colletotrichum capsici*. The infected plants show burnt appearance. These leaf spot diseases can be controlled by spraying Dithane M-45 (0.3%).

Inflorescence blight (*Alternaria zinnae*): Initially, minute brown spots appear, which enlarge into blotches. The spots show zonations or concentric circles (Target board pattern). Inflorescence gives a burnt appearance in severe cases. This disease can be controlled by spraying Dithane M-45 (0.3%).

Flower bud rot (*Alternaria dianthi*): This disease appears on inflorescence or young buds, which show discoloured appearance in initial stages. The buds show dry rotting in case of severe infection. Bud rot can be controlled by spraying Dithane M-45 (0.3%).

Powdery mildew (*Oidium* species, *Leveillula taurica*): Initially, whitish tiny spots appear on leaves. The entire plant shows white powdery appearance in severe cases. The disease can be controlled by the spray of Karathane (0.1%).

Viral diseases (Cucumber Mosaic Virus, Aster Yellow Virus): Infected plants show chlorotic streaking and yellow mosaic appearance. Infected plants get dwarfed. Roguing of diseased plants and spraying 0.25% Imidacloprid to control vector population prevent the spread of the disease.

SUGGESTED READING

1. Arora JS 1983 Marigold, in *Commercial Flowers,* (eds) TK Bose and LP Yadav, Kolkata: Naya Prokash Publications, 713-731 pages
2. Coats AH 1968 *Flowers and Their Histories*, McGraw Hill Book Co., New York,pp 242
3. Janakiram T 1997 Production technology of marigold. In: IS Yadav and ML Choudhary (ed) *Progressive Horticulture.* Sarpan Media House, Bangalore,pp 148-151
4. Negi SS and Raghava SPS 2000 Breeding of ornamentals. In: V L Chopra (ed) *Plant Breeding–Theory and Practice* (2nd Edition). Oxford and IBH Publishing Co. Pvt. Ltd., New Delhi, pp 279-320
5. Raghava SPS 2000 Marigold - Versatile crop with golden harvest, *Delhi Garden Mag.* **39** 14-15
6. Raghava SPS 1997 Pusa Narangi Gainda and Pusa Basanti Gainda, new marigolds, *Indian Horticulture* **43** : 31
7. Raghava SPS, Misra RL and Dhyani D 1987 Marigold - A popular commercial flower, *Delhi Garden Mag.* **26**: 25-29
8. Roy RK and Sharga AN 2000 Marigold- NBRI, *Information Bulletin* No. 7, pp 27
9. Sreekala C 2000 Production of F_1 hybrids in African marigold (*Tagetes erecta* L.) for higher content of carotenoids, Ph D thesis, IARI, New Delhi
10. Tomar BS and Singh KP 2009 Marigold for year round livelihood, *Indian Horticulture* **54**: 19-21.

GLADIOLUS

Devendra Dhyani

INTRODUCTION

Gladiolus is very popular as a cut flower, both with the consumer and the florist. It occupies a prime position in the floriculture industry among the flowering bulbous crops in the Indian markets and can be grown successfully both in the hills and plains. There are many forms, colours and colour combinations available in gladiolus. Therefore, they can be easily used in any of the numerous floral and bouquet arrangements, herbaceous borders, pots, rockeries, beddings and for cut flowers. The Roman and Greeks used gladiolus on important occasion more than 2000 years ago. The genus *Gladiolus*, belonging to the *Iridaceae* family has many species and the distribution of *Gladiolus* species is very wide throughout the temperate zone of the world. Most of the species are native to South and Central Africa and also occur around the Mediterr-anean region, Asia Minor, the Middle East, Turkey and Iran. There are more than 200 species of gladiolus reported in the literature world over and approximately only 23 species have been involved in hybridization so far to develop new hybrids (Anderton and Park 1989). No gladiolus species is known to be native to India (Misra and Singh 1989). Frequent hybridization and selection have replaced the old species and many of the wild species do not look like the gladioli we use as cutflowers now. The history of gladiolus cultivation reveals that it is known since 1578, as evidenced by a record in Lyte's Nievve Herball, first introduced into France and then spread to England, Germany, Holland and North America (Misra and Singh 1989).

In India, gladiolus was cultivated for the first time in the year 1863 by Charles Gray of Coonoor in his garden according to Firminger's Manual of Gardening in India, which was published in 1863 (Misra et al 2006). The attention of Indian researcher went on gladiolus only when Professor Harbhajan Singh brought 100 varieties in the Department of Botany of Indian Agricultural Research Institute, New Delhi in 1950. Late Shri Bajrang Bahadur Singh Bhadri, the then Governor of Himachal Pradesh, recorded the possibilities of gladiolus cultivation in Himachal

Pradesh and other hilly regions in a bulletin published in 1963 by Department of Agriculture Himachal Pradesh. He started gladiolus hybridization in 1957 and developed nearly 150 hybrids (Misra 1995 and Dhyani and Mukherjee 2008). Afterwards, National Botanical Research Institute, Lucknow; Indian Horticulture Research Institute, Bangalore; Institute of Himalayan Bioresource Technology, Palampur and other institutions started research work on gladiolus. It is now successfully grown in the hilly regions and plains of almost all the States of India and its area under cultivation is increasing gradually for cut flower production. The agroclimatic conditions of the North Indian hills and other hilly regions of the country are more congenial for gladiolus cultivation due to mild temperature, dust-free environment, abundant sunshine and prolonged growing period. Gladiolus grown in these areas produce better quality flower spikes with larger number of florets, better colour consistency, and greater cormel production. In these areas, time taken by cormels to attain flowering size corm is less and the supply of cut flowers to the main markets is possible during off-season when there is no production in the plains.

DESCRIPTION OF THE PLANTS

Gladiolus is an ornamental herbaceous plant, belonging to the family Iridaceae with chromosome number, n = 15. Most of the members of this genus are heteroploids having chromosome number ranging from 2n = 30 to 180 (Bhattacharjee and De, 2003). The plant produces corm at the base under the soil surface, which consists of compactly arranged nodes. The corm is surrounded by 4-6 dry scales, which are the lower parts of the leaves formed during the earlier growth season and these scales protect the new buds present on the corms.

After planting of corm, 6-8 healthy leaves develop successively before the appearance of flower spikes. When the third leaf appears, bud initiation is in full progress and a new corm is formed -on old mother corm and it develops a number of contractile roots which assume the function of the old corm roots. This period is known as the critical period in which lack of soil moisture, nutrients, light intensity and temperature can increase the chance of flower desiccation and can arrest the growth, thereby, ultimately affecting the development of flower spikes and their quality. Depending upon the size of corms and cultivars, one or more buds sprout from the corms. However, formation of flower on the plants depends on corm size, light intensity, planting density, cultural practices and other environmental factors. Two types of roots are formed by the gladiolus plants. One type of roots arises from the base soon after planting of corm or cormel and the other type of roots is the contractile root, which arises from the base of the newly formed daughter corm.

CLASSIFICATION

The classification system of gladiolus varies in different parts of the world. Variations exhibited by varieties and species in their flower characters, colour, vegetative growth and time taken to flower are many and are classified accordingly in different countries. However, the American system of classification developed by the North American Gladiolus Council (NAGC) based on flower characters is more scientific and comprehensive and is now accepted and used widely. The following three systems are generally used for classification of gladiolus cultivars.

Table 12. Classification of gladiolus based on North American Gladiolus Council

Designation	Flower size (cm)
i. Miniature	< 6.4
ii. Small	> 6.4-8.9
iii. Decorative	> 8.9-11.4
iv. Standard or large	>11.4-14.0
v. Giant	>14

Table 13. Classification based on time taken to flower after corm planting

i. Early flowering	Blooming within 60 days
ii. Mid Flowering	Blooming within 60-90 days
iii. Late flowering	Blooming after 90 days

Table 14. Classification based on plant height

i. Dwarf	Plants attaining height up to 50 cm
ii. Medium	Plants attaining height from 50-100 cm
iii. Tall	Plants taller than 100 cm

CULTIVATION

Selection of Site

Gladiolus needs full Sun throughout the growing season and can be grown successfully in outdoor conditions. In the areas experiencing cool climate with chances of frost and heavy rains during the growing period, quality flowers can be grown under polyhouse conditions. While selecting the site for growing gladiolus, planting time and the climatic conditions of the place should be taken into consideration. A place having high wind intensity during the crop season or shady places should be avoided for raising gladiolus.

Soil and Soil Preparation

Gladiolus can be grown in a wide range of soils, provided the structure and drainage of the soil are good. However, best spikes are produced when plants are grown in sandy loam soil. Heavy clay soil, with poor drainage restricts the growth of plants due to poor root development and promotes disease incidence due to excessive soil moisture. However, heavy soils can be made cultivable by mixing a thick layer of sand along with well rotten manures. Soil having a pH range between 5.5 and 7 is suitable for cultivation. Gladiolus needs well prepared soil for both flower and cormel production. The first ploughing of the soil at selected site should be done to a depth of 30-40 cm at least two months ahead of planting. Second ploughing should be done about 2-3 week before corm planting, followed by rolling to achieve fine tilth. Weeds should be removed meticulously and beds should be prepared for planting as per the soil and climatic conditions of the location.

Soil Treatment

Planting of gladiolus corms in fresh soil is always beneficial for quality flower and planting material production, and for avoiding damage caused by the soil borne diseases. Crop rotation of 5-6 years is also beneficial to minimize disease incidence, if fresh plot is not available. Soil used in previous years of the cultivation of gladiolus, Iris and Freesia must be disinfected before use. Soil sterilization should be done at least one month prior to corm planting and if required sand and manure should be added to the soil before the chemical treatment. Soil sterilization with 2% formaldehyde solution is effective to control fungal pathogens present in the soil. About 30-40 cm deep soil should be dug out from the beds and the bottom soil should be drenched with formaldehyde solution quickly. The trench should be filled and leveled with the removed soil. The bed should then be quickly covered with polyethylene sheets for at least 15 days. Soil should be covered in such a way that the fumes do not escape. After 15 days of sterilization process the polyethylene sheets should be removed and the soil should be dug again and irrigated thoroughly before planting.

Selection of Cultivars

Selection of cultivars for cut flower production depends on various factors such as suitability of the variety for selected site, plant height, size of flowers, preferred colour choice by markets, purpose of growing and availability of planting material. Mainly pure white, bright red, orange, yellow and pink colours are preferred and have good demand in the markets compared to other coloures and bi-coloured varieties. Some of the important cultivars of gladiolus are Almost Heaven, Applause, Bewitched, Black Swan, Bonaire, Carved Ivory, Dark Victory, Eurovision, Excellence, Fidelio, Forest Fire, Friendship, Hunting Song, Jessica, Jester, Jester Gold, Joyful, Lady Friend, Nova Lux, Oscar, Peter Pears, Pink Dream, Pink Perfection, Priscilla, Red Beauty, Red Lace, Rose Supreme, Victor Borge, Visionary, White Friendship, White Goddess, White Prosperity, White Cloud and Wind Song.

Selection of Corms

Selection of good quality planting material is crucial for cut flower and planting material production. Climatic conditions and cultivation practices to be followed are also important considerations. A high crowned, young, fleshy corm is better than a larger and flat old corm as a young corm has better vigour than the old one. The age of the corm can be determined by examination of the base or root scar. A young corm has a small root scar where as an old corm has a large one. Selected corms should be free from diseases and injuries.

Nutrition

Different doses of fertilizers have been suggested in the literature by different authors for gladiolus cultivation. Gladiolus do not require exessive fertilization. It is sensitive to high salt content in soil and therefore, heavy fertilization has a retarding influence on root growth and flowering. Fertilizer requirements vary with type of soils, fertilizers and salt content of the soil and climatic conditions. Depending upon the prevalent conditions, nitrogen requirement may vary from 200-500 kg per hectare, which should be applied in two splits. 200 kg each of phosphorus and potash per hectares is optimum. To improve the structure of soil, mix well decomposed FYM @ 10 kg per square metre on soil surface before planting. Application of un-decomposed FYM should be avoided because of its unfavourable influence on the growth. In the hills, following regime is sifficient to harvest a good crop:

- Application of 50 gm of single super phosphate and 25 gm of Muriate of potash per square metre area at the time of soil preparation
- Application of calcium ammonium nitrate (CAN) @ 40 gm per square metre from the emergence of plants up to 3 leaf stage
- Another 40 gm after one month of the first application

Watering

Gladiolus corms require sufficient moisture in the soil after planting for quick and undisturbed rooting. Therefore, watering of the field should be done a few days prior to planting, in such a way, that the corms can be planted when the soil is moist but not excessively wet. If sufficient moisture is present at the time of planting, little water is required till the emergence of plants. Thereafter, sufficient moisture should be maintained in the field throughout the growth period. The most critical period for water requirements begins at the time when the third leaf is visible and ends when the seventh leaf appears. In this period flower initiation is in progress, the roots of the old corm die and the contractile roots of the new corms are initiated. Therefore, extra attention is required in this period for proper water supply to the crop.

Planting Time

In the North Indian plains, gladiolus is planted during winter season in the months of October-November. In the hilly regions, it is generally planted in the month of February and onwards. Planting time may vary from place to place as per the climatic conditions and altitude of the place.

Planting Method

Method of gladiolus planting depends upon the location, soil and climatic conditions and irrigation system to be employed. Depending upon the conditions, it can be planted in flat beds, raised beds and ridges. In light soils such as sandy and sandy loams, it is advisable to plant them in ridges or in 1 m wide raised beds. In the hilly regions having high rain fall during crop season, planting should be done in raised beds and ridges to facilitate the excess water to drain and hold the plant intact in the field.

Corm Size and Planting Density

The flower quality of the crop depends mainly upon the size of the planted corm. A large size corm will produce taller and stronger plants with more uniform flowering, heavier spikes and in shorter period as compared to a smaller sized corm. Planting density mainly depends upon the size of corms and it is important for firmness of the plant and the quality of flower. Large number of corms planted per unit area make the plants long and limp, and the spikes become short due to light deficiency. Planting of gladiolus is done for two purposes i.e for cut flower and for planting material production. Corms having a diameter of 2.5 cm and above should be used for cut flower production, while small corms and cormels are generally used for planting material production. Appropriate number of corms need to be planted per unit area as per their size to get a crop of good quality. Gladiolus corm size is generally measured and graded worldwide on the basis of its circumference, in centimetres. The following table indicates the size of corm and its density per gross square metre area (Buschman):

Table 15. Number of corms per unit area

Corm size (cm)	Number of corms per gross m^2
6-8	60-80
8-10	50-70
10-12	50-70
12-14	30-60
14 and upwards	30-60

Depth of Planting

The depth of planting of corm in soil mainly depends on the size of corm, type of soil and time of planting. Considering these factors and purpose of planting, corms and cormels can be planted at 5-15 cm depth (Chattopadhyay, 1995). Corm planting should be done in such a way that 5-10 cm soil layer covers its top. Spring planting can be done at 5-10 cm depth and summer planting at 10-15 cm depth. Deep planting in summer protects the bulbs from high and fluctuating temperatures, and restricts toppling of the gladiolus plants in sandy soils.

Mulching

Mulching is not generally practised in the cultivation of gladiolus flowers. However, it has some beneficial effects. The soil temperature can be kept low by using mulch of straw or pine needles during summer planting and it can maintain higher temperature during winter. In the areas likely to experience frost during winter, heavy mulch can avoid the risk of frost damage to the crop. Mulch also has a highly positive effect on the maintenance of the soil structure in the area of heavy rainfall.

Light

The gladiolus plants need plenty of sunlight throughout the growing season. High light intensities and long day conditions are ideal for proper growth of the plant, quality of flower and planting material production. Low light conditions during 3-6 leaf stage adversely affect the height of flower spikes, and number and quality of florets in the spikes. Suitable light conditions are also required for utilization of nutrient reserves of planted corms through the assimilation process, which are essential for good growth and flowering of the gladiolus plants.

Temperature

A temperature between 10-25°C is most suitable to obtain optimum growth of gladiolus plants. It can be grown above 25°C and can tolerate temperature up to 40°C temporarily if the soil moisture condition is optimum and the relative humidity is high. Plant growth and development is arrested when the average temperature goes lower than 10°C. Gladiolus plant can not tolerate temperature below freezing point. Interruptions in growth may occur due to prolonged cold weather conditions. There is a significant relationship between incident light and favourable temperature with the period when flower develops in the gladiolus plant. The period of flower development lies between the time third leaf becomes visible, and the time the sixth or seventh leaf appears. During this period, excessive average temperature in combination with deficient light may cause flower desiccation. In the last part of this period, adverse temperature and light conditions may also affect the number of flower buds per spike.

Dormancy of Corms

Dormancy in gladiolus corms and cormels is a period when growth and development are temporarily stopped. After harvesting the corms, they will not sprout if sown immediately, even under favourable growing conditions. They will not sprout for 3-4 months due to the dormancy or rest period of the corms. Dormancy period varies from cultivar to cultivar and between corms and cormels, and depends on the cultivar and growing conditions. In cormels, the dormancy is more prolonged than in corms. This dormancy can be broken by storing the corms in the cold storage. In the plains, corms are stored at 4°C for a period of 2-3 months to break the dormancy. Whereas, in the hill regions, they can be stored at room temperature in a well ventilated place for the same period during winter months.

Weed Control

Weeds affect the quality and yield of flowers and corms in gladiolus. Weeds need to be controlled before the crop comes up. Perennial weeds can be controlled by spraying Glyphosate or by rounding up well before the planting at the rate of 6 litres per ha. Gramoxone, 6 litres per ha is useful to control annual weeds. Hand removal of weeds from time to time is necessary for proper growth of the crop as it minimizes the competition for nutrients between the crop and the weeds. Weed control after crop emergence to the seventh visible leaf stage promotes healthy crop as it is the most critical period for water and nutrient requirement, and development of flower and new roots from the planted corms. The herbicides should be sprayed between the rows in the evening on moist and compact soil.

CONTROL OF INSECT PESTS AND DISEASES

Gladiolus corms and plants are infected by various fungal and bacterial diseases and infested by insects which cause substantial loss to the crop and planting materials. Clean cultural practices, preventive measures for pest and disease control with regular application of plant protection chemicals are essential for quality flower and planting material production. The most common insect pests and diseases are described below:

Insects

Aphids

Aphid is one of the common insects that infests gladiolus crop in the warm months. They suck the sap from very tender shoots and produce inferior quality flowers. They also spread viruses in the crop. Spraying of systemic and contact insecticides such as Malathion, Rogor and Metasystox @ of 2 ml per litre of water at an interval of 15-20 days during the early part of the growing season can control the aphids.

Thrips

Thrips feed on the tender epidermal layer of leaves. The infested leaf surfaces have unnatural whitish grey colour. Thrips also suck the sap from the flowers and buds, which become deformed and fail to open. The young thrips are light yellow in colour. They feed under the leaf sheaths and appear mostly in cloudy weather. Infested flowers are discoloured and spotted. The foliage of infested plants has many whitish spots and later the leaves turn yellow and dry out. Spraying of Methyl parathion (Metacid-50) @1ml per litre or Endosulfan 35 EC (Thiodan) @ 2ml per litre of water can effectively control thrips.

Cut Worms

They attack different parts of the plant and cut the basal region at a young stage. Cut worms hide in the soil surrounding the plants during daytime and appear when it is dark. It can be controlled by fumigation of infested soil and fortnightly spray of 0.2 % methyl parathion.

Mites

Several species of mites attack gladiolus plants during warm, humid and shady conditions and nourish on the leaves. White specks appear on leaves, which ultimately cause bronze or silvery appearance to the affected plants. Timely spraying of suitable miticide can control it.

FUNGAL DISEASES

Plants and corms of gladiolus that have been infected by various types of fungal diseases in the field and during storage, cause substantial loss of planting material and flower spikes. Some of the important diseases are described below:

Botrytis Dry Rot

Botrytis is caused by the fungus *Botrytis gladiolorum* and it is a highly destructive, soil borne and most common cause of leaf and flower spot of gladiolus infections. It is also known as soft corm rot, neck rot, and grey mould. The symptoms on the corm consist of tiny black-brown spots at the top of the corm around stalk base and brown strains of variable shape and size on the surface. In cases of severe infection, the corms become soft and reddish brown in colour and when the corms dry up, they shrivel and become hard. On plants, the infection can occur at soil level and the plants become yellow. The symptoms on the leaves consist initially of light and later dark brown round spots. Later in the season, large and dead grey-brown patches appear on the tissue. Germinating spores may cause colourless watery spots on the flowers. Sclerotia can be formed on all parts of the plant, as well as on the corms. They are variable in size, flat and dark coloured, and can spread the disease in the next season. The corm infections arise particularly under high moisture conditions during storage of corms, while infection of leaf and stalk at soil level is caused under cool and moist conditions. To control the disease during storage of corms, the relative humidity is kept low and good air circulation is maintained. Corms need to be treated with suitable fungicides before planting. Cultural control measures should also be practiced such as sufficient crop rotation and avoidance of dense planting. The crop should be regularly sprayed with fungicides at an interval of 15-20 days to prevent disease spread.

Fusarium Rot

The fungus *Fusarium oxysporum* causes this disease and it is commonly known as vascular disease. This is the most serious storage disease of gladiolus corms. The fungus may be present on the corms when they are lifted from the soil and rot develops rapidly in storage, especially if the corms have not been dried thoroughly and cured before being stored. During the storage, the spores of this fungus are spread easily by air circulation. Infection from the soil is also possible. Severely infected corms do not come up and partially infected corms produce weak and curved plants with drying of leaves. Initially, only the leaf tips of these plants are yellow and afterwards the whole plant turns brownish yellow and dies. The corm base turns grey to dark brown and this discolouration continues in the central part of the corm and spreads to the sides. For effective control of this disease, remove the plants from the soil as soon as the symptoms appear. Use of disease-free planting materials is recommended. Dusting with 10-20% benomyl or TBZ (thiabendazole) before planting and after harvesting of corms is found very effective to control the disease.

Dry Rot

Dry rot is caused by the fungus *Stromantinia gladioli* or *Sclerotinia gladioli*. It is also known as neck rot or root rot and causes yellowing or death of the foliage. Dry rot of corms caused by *Sclerotinia gladioli* is seen on the storage corms as small, dark spots or lesions. In warmer areas, dry rot occurs during winter cultivation at lower temperature. The fungus grows below the ground from an infected plant and spreads to surrounding healthy plants and keeps the soil infected for long time. At first the plants show a good growth, but after some time the foliage turns yellow starting from the the outer leaves. The spathe leaves of affected plants are brown to black and are covered with many small, round, black sclerotia, which develop on the leaves and husk of the corm, which carry the disease to the next season. The spots on the corms are usually circular and sunken, but have no gummy surface like those of scab. The brown or black lesion that appears on the corms may destroy the whole corm, especially in moist storage conditions. The infected corms ultimately become hard and black. It is a wide spread disease of gladiolus infecting plants in field and corms in storage. Removal of the infected plants with the corms and using healthy planting material in a new area is recommended.

Curvularia Blight

It is caused by *Curvularia trofolii* var. *gladioli* and infects all plant parts including leaves, stems, flowers and roots. It is a serious soil-borne disease of gladiolus that limits the quality flower spike production. This pathogen survives in the soil for three or more years. Oval-shaped tan or dark brown spots appear on the leaves and stems. When the stems are heavily infected, florets fail to open. Dithane M-45 and Miltox each @ 0.2% sprayed at fortnightly interval can control the disease.

Root Rot

The rot is caused as a result of infection by the *Pythium* fungus, the root lesion nematode or other non identified fungi and bacteria. The infected plant retards shortly after coming up until flowering in patches. It flowers later or not at all and sometimes die down prematurely. Soil disinfection is recommended for the control of the disease.

Rust

This disease is caused by the fungus *Uromyces transversalis* causing rust on the leaves. Yellowish spots develop on both sides of the leaves of the infected plant. These spots are circular at first, but enlarge in width afterwards. They burst after some time and orange-yellow spores are released. The orange-yellow spores are responsible for spreading the disease. The disease occurs in hot and humid conditions. The plants need to be treated with orycarboxin (5 g Plantvax 75% per 100 square metre) or benodanil (10 g of Calirus 50% per 100 square metre) every week after the disease is identified.

BACTERIAL DISEASES

Bacterial Blight

Bacterial blight is caused by *Xanthomonas gummisudans*, which results in neck rot and spots on leaves. This disease is serious especially in wet weather or on plants grown in poorly drained soils (Pascal and Pirone, 1978). At initial stages, irregular water-soaked spots appear on the leaves which dry out and turn brown at later stages. Gradually, the whole leaf shows disease symptoms and dies. This disease infects the plants seriously in wet weather and poorly drained soil conditions.

Scab

It is a common disease of gladiolus and is also known as neck rot and stem rot, caused by *Pseudomonas marginata*. Pale yellow, water-soaked circular lesions appear on the corms, which turn brown or black, become sunken with raised bristle margins that are scab like and exude a gummy substance. Early symptoms on the leaves are reddish raised specks near leaf base. The neck or fleshy basal parts rot as the disease progresses and leaves fall. For the control of bacterial blight and scab, discard the bulb and plants that show the symptoms as early as possible and use disease-free corms for planting.

Virus

Many viruses are known that infect gladiolus plants. Viruses such as tobacco ringspot virus, bean yellow mosaic virus, tomato ring spot virus and cucumber mosaic virus are common. Most of the viruses of gladiolus are transmitted by aphids or leafhoppers. Tobacco ringspot virus can be transmitted through tools during the flower and corm harvesting. Since the viruses are carried over from one season to another in the corms, infected plants should be discarded completely from the field. Regular spray of suitable chemicals for controlling insects can reduce the occurrence of the viruses.

DURATION OF CULTIVATION

Gladiolus needs at list 6-7 months for cultivation from planting of corms to harvesting of corms. It mainly depends upon the type of varieties, corm size and average cultivation temperatures of the growing area. Generally, duration of gladiolus cultivation is understood to be the total number of days required from planting of corms until flowering.

According to the length of this cultivation period, cultivars can be listed as short, medium and long duration. A variation in the average cultivation temperature also causes a difference in the duration of cultivation. The average cultivation temperature and duration of cultivation are given in the table.

Table 16. Average cultivation temperature and duration of cultivation

Average cultivation temperature (°C)	Duration of cultivation (days)
12°C	110-120
15°C	90-100
20°C	70-80
25°C	60-70

HARVESTING AND STORAGE OF FLOWERS

Flower spikes of gladiolus should be harvested at right stage to realize its proper value. For distant markets, the flowers can be picked as soon as the bottom flower bud of the spike starts colouring. If the flower bud is allowed to open, it may be damaged during processing and transport. If the flower spikes are harvested too early i.e. when the bottom flower bud shows no colour, it may not flower properly. Flowers should be harvested with a sharp knife in the morning hours; leaving 4-5 leaves on the plant and immersed immediately in buckets having clean water to maintain the turgidity. Open flower spikes can be used for local markets, bouquet preparation and home decorations. Flowers can be stored at a temperature of 2-5°C under dry conditions for short durations to maintain the durability of the flowers for the consumers. Flowers after harvesting should be cleaned, stored and transported in an upright position. This will prevent the tops of the spikes from bending and getting crooked.

Grading of Spikes

Gladiolus spikes are mainly graded for commercial purposes on the basis of spike length and number of flowers per spike. Apart from their overall quality, spikes should be strong and straight. This system of grading is given below:

Tabel 17. Grading system

Grade	Spike length (cm)	Minimum number of florets
Fancy	> 107	16
Special	> 9 - 107	15
Standard	> 81-96	12
Utility	< 81	10

(Wilfret 1980)

PROPAGATION

Gladiolus can be propagated by sexual as well as asexual methods depending upon the purposes. Sexual propagation is used to raise new cultivars by hybridization programme followed by selection of promising genotypes, whereas a cultivar is propagated by corms and cormel. In new growing season, new corms are produced from the planted mother corms and many cormels are produced at the tip of branched stolons that develop from the buds located at the base of the new corm. Commercial method for maintaining stock and supply of flowering size corms is by planting cormels every year. It is the quickest method for propagation of gladiolus.

HARVESTING AND STORAGE OF CORMS

Gladiolus plants generally mature after 60-90 days of flowering, which depends upon the variety and place of growth. Crop maturity and harvestability can be identified by the browning of leaves and drying of the plants, which is completed in between last week of April to first week of May, in Northern Indian plains. Depending upon the planting time, height and weather conditions in the hills, corms can be harvested between the months of September-November. Once the browning starts in the plants, no irrigation should be given to the crop as it may cause rotting of the corms. The corms should be lifted by spade or Khurpa in such a way that the corms do not get any mechanical injury. Harvesting of corms at the right stage is most important to obtain quality planting material for the subsequent cultivation.

After harvesting, corm and cormels should be washed in running water, dried under shade and graded. Harvested corms should be disinfected with suitable fungicides before storage and again immediately before planting. The corms can be dipped in a solution of 0.2% captan and 0.2% benomyl for at least an hour to reduce disease incidence. Treated corms and cormels can be kept in the plastic trays and gunny bags and stored in a dark and well ventilated room or cold storage. The ideal storage conditions for corms and cormels are 3-7°C temperature and 75% relative humidity.

YIELD

Yield of gladiolus flower spikes and corms per unit area varies according to the cultivar, growing conditions, corm size, planting density and adopted cultural practices. Generally, one plant produces a single marketable spike and plantable size of corm. High planting density with suitable management practices have been found to give higher yield per unit area. Approximate yield of flower spike would be around 2-2.5 lakh per hectare in open cultivation (Mukherjee, 2008).

REFERENCES

1. Anderton EW and Park R 1989 Growing Gladioli. Publishers Christopher Helm Ltd. Imperial House, 21-25 North Street Bromley, Kent BR1 1SD.
2. Bhattacharjee SK and De LC 2003 Gladiolus, in *Advanced commercial floriculture* (Volume 1), Jaipur (Rajasthan) India: Aavishkar Publishers, Distributors, 309-321 pages
3. Buschman JCM *Gladiolus as cut flower in subtropical and tropical regions*. International Flowerbulb Center, Parklaan 5, PO Box 172, 2180 AD Hillegom-Holland.
4. Chattopadhyay TK 1995 Cultural Requirements of Gladiolus, in *Advances in Horticulture* (Volume 12), (eds) KL Chadha and SK Bhattacharjee, New Delhi: Malhotra Publishing House, 725-742 pages
5. Dhyani D and Mukherjee D 2003 Gladiolus Ki Sankar Prajatiyon Ka Vikash, *Bhartiya Vigyanik evam Audyogik Anusandhan Patrika* **11** 180-183
6. Misra RL and Singh B 1989 Gladiolus, in *Commercial Floriculture* (eds) TK Bose and LP Yadav, Calcutta: Naya Prokash, 267-363 pages
7. Misra RL, Hussan Sakkeer CT and Misra Sanyat 2006 Gladiolus, in *Advances in Ornamental Horticulture* (Volume 3), (ed) SK Bhattacharjee, Jaipur (Rajasthan) India: Pointers Publishers, 88-106 pages
8. Misra RL 1995 Genetic resources of Gladiolus, in *Advances in Horticulture* (Volume 12), (eds) KL Chadha and SK Bhattacharjee, New Delhi: Malhotra publishing House, 193 pages
9. Mukherjee D 2008 Gladiolus, in *Specialty Cut Flowers Production Technologies,* Kolkata, India: Naya Udyog, 521-586 pages
10. Pascal P Pirone 1978 Diseases and Pest of Ornamental Plants, New York: John Wiley & Sons, 277-282 pages.

TUBEROSE

B.K. Banerji

INTRODUCTION

The Sanskrit name for tuberose is 'Rajnigandha' which means 'fragnance of the night'. Botanically, *Polianthes tuberose* was already domesticated in Mexico at the time of the Spanish conquest. No forms of tuberose have ever been discovered growing in the wild. Tuberose is an important bulbous ornamental of tropical and subtropical areas and is grown as a garden plant for its fragrant flowers and is commercially cultivated for extraction of its highly valued natural flower oil. In India, it is grown extensively in the eastern part of the country, but is also cultivated in northern plains and parts of the south. When grown only for flower production, the climate must be warm and humid. Profuse flowering can be obtained throughout the year where the climate is mild.

SOIL AND SOIL BED PREPARATION

The plant can be successfully grown on a wide range of soils including those inflicted with mild acidity or alkalinity. It is very sensitive to water-logging. Sandy and sandy loam soils in pH range of 6.5-7.5 with proper aeration and drainage are considered best for tuberose cultivation.

RECOMMENDED VARIETIES AND THEIR SALIENT FEATURES

The main varieties are 'single' with one row of corolla segments and 'semi-double' with two to three rows of corolla segments. The varieties of single tuberose commercially grown in different states are Calcutta Single, Calcutta Double, Hyderabad Single, Hyderabad Double, Phule Rajni, Mexican Single and Sikkim Selection. Four new cultivars of tuberose have been developed at the Indian Institute of Horticultural Research, Bangalore viz. Prajwal, Suvashini, Vaibhav and Shringar. Two mutants of tuberose induced by gamma rays have been released from National Botanical Research Institute, Lucknow in the year 1974. The details of the important cultivars and their characters are given in Table 18.

Table 18. Vegetative, floral characters and uses of tuberose cultivar

Name of cultivar	Bud colour	Foliage and Flower colour; Spike, Seed setting and Concrete characteristics	Use of flowers
SINGLES			
Calcutta Single	White with a tinge of pink red	Foliage green, flowers white, Spike grows straight	1.Extensively used as cut flower and loose flower 2. Extraction of essential oil
Mexican Single	White	Foliage green, flowers greenish white, highest concrete recovery 0.135%, bears seed	1. Extensively used as cut flower and loose flower 2. Extraction of essential oil
Shringar (Mexican single X Pearl double) IIHR, Bangalore 1994	White	Foliage green, flowers pinkish white, 36% higher flower yield, concrete recovery 0.133%, Produces seeds under favourable conditions	1. Extensively used as cut flower and loose flower 2. Ideal for pot culture 3. Suitable for land scaping
Prajwal IIHR, Bangalore 2001	White	Foliage green, flowers pinkish borne on a tall and stiff spike, flowers are large in size with heavy weight	1. Extensively used as cut flower and loose flower 2. Extensively used in perfume industry
Arka Nirantar IIHR, Bangalore 2010	White	Foliage green, flowers pinkish, early blooming, higher spike yield, profuse blooming	1. Ideal for loose flower 2. Used as cut flower and loose flower
Phule Rajni (Local single X Shringar) MPKV, Rahauri 2008	White	Foliage green, flowers greenish white, Spike straight with 52 florets (Range 46-58 florets), Rachis length is more than 50% of the spike length	1. Extensively used as cut flower and loose flower
Rajat Rekha (Gamma ray induced mutant) NBRI, Lucknow 1974	White	Foliage variegated, flowers white, Spike straight, silvery white streaks are very prominent along the margin of the blade	1. Ideal as a potted plant and can be used for decoration of lawn 2. Used as cut flower and loose flower
Double			
Pearl Double	Tinged with pinkish red	Foliage green, flowers dull white, incomplete opening of the flowers, blooms are sterile, No seed formation	1. Used as cut flower

Name of cultivar	Bud colour	Foliage and Flower colour; Spike, Seed setting and Concrete characteristics	Use of flowers
Suvashini (Mexican single X Pearl double IIHR, Bangalore 1994	Pinkish white	Foliage green, flowers white, big and bold, all florets of the spike blooms	1. Used as cut flower 2.Spikes are best suited for cut flower
Vaibhav IIHR, Bangalore 2002	Greenish white	Foliage green, flowers white, Plant attains medium height	1. Best suited for cut flower
Calcutta Double	Tinged with pinkish red	Foliage green, flowers white with a tinge of pink-red	1. Used as cut flower
Swarn Rekha (Gamma ray induced mutant) NBRI, Lucknow 1974	Creamish white	Variegated foliage, flowers white and double, Leaves with golden yellow streak along the margin	1. Ideal as a potted plant and can be used for decoration of lawn 2. Used as cut flower and loose flower

PROPAGATION

Tuberose is propagated through seeds and bulbs. Vegetative propagation is preferred because the resultant plants produce better flowers within a short period of planting. Bulbs of 2-3 cm diameter are suitable for propagation. Planting of freshly harvested bulbs produces smaller number and therefore, bulbs should be stored for a month or more before planting to ensure better production of flowers. Planting larger-sized bulbs gives early flowering and higher yields. The bulbs should be planted 4-5 cm deep in beds and soil moisture should be maintained till onset of monsoon. Old and dead roots at the base of the planted bulb should be removed. The pips, when planted 12-15 cm apart in drills and properly tended, can yield blooming bulbs for the following year.

High light intensity is required for better bulb and flower production. For higher bulb production, tuberose must be grown under full Sun. It has been reported that exposure to a day length of approximately 16 hours promoted the vegetative growth of plant and enhanced the emergence of spike by 10 days. Better bulb growth has been observed in large pots (25 to 30 cm). Since plants grown in pots tend to be warmer than those grown in the ground, they often bloom a few weeks earlier.

Tuberose is a tropical plant and grows well round the year under moderate climatic conditions like that obtained in Bangalore. However, it is well adapted to subtropical conditions (North Indian plains) and grows profusely from April-November. It becomes dormant with the advent of winter season, as evident by the cessation of vegetative growth. High temperatures (40°C) as well as low temperatures (10°C) reduce the length of spike and also weight and quality of harvested flowers.

CULTIVATION

Planting Season and Spacing

Tuberose can be commercially grown throughout the year but highest yield is obtained from February-October. However, best time for planting is from second to last week of March in the plains and April-May in the hilly areas.

Nutrition

The nutritional requirement for tuberose crop varies with several factors such as climatic conditions, types of soil, cultivars planted and spacing followed. It was found that nitrogen is more vital than phosphorus or potassium in respect to yield and quality of flowers and bulb production in tuberose. During preparation of the soil, a basal application of FYM at the rate of 12-15 tonnes per ha is recommended for better growth and flowering. Besides the application of FYM, tuberose requires balanced application of major and micro nutrients. A fertilizer mixture containing 6 g of urea, 16 gms each of single super phosphate (SSP) and potash per square metre gives satisfactory growth and flowering. This mixture should be applied twice. The first dose should be applied before planting and the second dose, four weeks after sprouting of the bulb. Foliar feeding with K and N produces greater number of flowers of better quality. Most of the nutrients become available to the plants in the pH range of 6.5-7.5 when soil has good aeration.

Land is prepared by ploughing two or three times. FYM @ 30 tonnes per hectare is mixed well with soil. The bulbs, preferably those of size 2-5 cm or above, are planted at a depth of 7-10 cm, with a spacing of 20 x 25 cm. A fertilizer dose of N, P_2O_5 and K_2O in the ratio 100:50:50 kg per hectare is recommended. Of these, half amount of N and whole amount of P_2O_5 and K_2O should be applied at the time of planting. Remaining amount of N should be applied when the flower spikes start to appear. A heavy irrigation once in 5-10 days is necessary depending upon the weather conditions. The peak flowering is between June and October.

Watering

Soil moisture is an important factor that affects the growth, flowering and bulb yield of tuberose. The field should be irrigated before planting of bulbs and irrigation should be avoided until the sprouting of bulbs. The crop should be irrigated at weekly intervals in the absence of rainfall. Mulching reduces the water requirement by minimizing the evaporation losses. However, irrigation should be avoided at the maturity stage of bulbs during December-January.

Mulching is primarily carried out for conserving soil moisture and checking weeds in the field. Black polythene and gliricidia leaves are most suitable mulches for tuberose cultivation. About 5 kg green leaves of gliricidia are sufficient for 1x 1 m plot to harvest a bumper yield and quality of flowers and spikes. Paddy straw can also be used for mulching to conserve soil moisture and to reduce the weed population. Basically, it needs warmth, sunshine, well drained soil, even moisture, and at least a month and a half of good growing conditions.

Weeding

Weeding is carried out fortnightly, especially during the initial stages of bulb sprouting and growth of plants. Hand weeding is eco-friendly but expensive.

INSECT, PESTS AND DISEASE MANAGEMENT

Incidence of various types of diseases on tuberose has been reported from different states of our country. Tuberose crop is attacked by various bacterial, fungal and viral diseases. Slugs and grasshoppers, foliar nematodes and root knot nematode also attack tuberose crop and cause severe damage. Pests which damage the crops include aphids, thrips, red spider mites, weevils, slugs and grasshoppers.

Bacterial Disease

Flower Bud Rot

Flower bud rot is a bacterial disease caused by Erwinia species. The disease appears mainly on young flower buds and results in dry rotting with brown scorched necrotic discolouration of peduncles. In the advanced stage, buds become shrivel and dry. This disease is mainly spread by thrips. Destroy or burn the infected plant debris to check further infection and adopt remedial measures to control the thrips.

Control:

1. Spraying of Rogor (0.1%) twice at a weekly interval is effective.
2. Destroy or burn the infected plant debris to check further infection.

Fungal Diseases

Flower Blight

The disease is caused by a fungus known as *Botrytis elliptica*. The disease appears in winter season. The flower blight is more common in double flower varieties of tuberose. The symptoms are visible on leaves and stalks as light brown lesion which starts from the margin of petals and gradually enlarges and ultimately causes blighting of the entire inflorescence. The infected spikes become thin and weak and drop down.

Control:

1. The disease can be controlled by spraying Carbendazim.
2. Application of nitrogenous fertilizers such as urea, ammonium nitrate and ammonium bicarbonate reduce the disease incidence.

Basal Rot

It is a widely spread and destructive fungal disease. In diseased plants, flaccidity and drooping of the leaves are clearly visible. It is a soil borne disease caused by the fungus, *Sclerotium rolfsii*. It can be identified with symptoms such as appearance of prominent coarse mycelia masses on the leaf surface at or near the soil level. The infection occurs first on old senescent leaves that are in touch with wet soil. The fungus causes maximum damage to crop at temperature ranging from 25°C-35°C. Presence of nematodes aggravates the disease. Infected leaves loose green colour due to rotting, which extends to the whole leaf and detaches the affected leaves from the plant. The fungus mainly attacks on the roots and the infection gradually spreads upwards through the collar portion of the stem. The tuber and roots show rotting symptoms, which are clearly visible in the form of thick cottony growth of the fungus on the rotten stem and on the petiole at the soil level. The infected plants should be burnt immediately to check further infection.

Control:

1. Trenching the soil with Zineb is effective to control the disease.
2. Application of ammonia fertilizers such as urea, ammonium nitrate and ammonium bicarbonate reduce the disease incidence.
3. Copper fungicides or Benzimidazoles are helpful and used by farmers to control the disease.

Sclerotium Fungus

Sclerotium fungus attacks the leaves and flower stalks at ground level, which causes defoliation and toppling of spikes.

Control:

Disease can be controlled by drenching of soil around the plant with fungicides.

Viral Disease

Poty Virus of Polyanthus

A poty virus of *Polyanthes tuberosa* has been reported from China, New Zealand and Taiwan. Infected tuberose leaf exhibits mottling symptoms. The causal organism is mild mosaic virus (TuMMV). It has also been reported from Himachal Pradesh.

Control:

1. Use of disease free plant material for cultivation is recommended.
2. The infected plants should be rooted out and burnt immediately to check further infection.

Insect-Pests

Slugs and Grasshoppers

Slugs and grasshoppers attack the tuberose plant. They feed on the leaves and cause damage to the plant.

Control:

Spray of 0.1% Malathion and carbaryl is effective against these pests.

Foliar Nematode

The damage caused by nematode in commercial cultivation of tuberose is severe in West Bengal. The disease was reported from Ranaghat area of Nadia district of West Bengal. The nematode has recently spread to Orissa presumably through bulbs. The primary causal organism is *Aphelenchoides besseye.* It is now a major limiting factor for cultivation of tuberose in Ranaghat, Haringhata and Banskera area of West Bengal. 'Calcutta single', a cultivar of tuberose, is more vulnerable to *A. besseyl* than the 'Calcutta double'.

The main symptom appears on spikes, where malformed flowers can be easily detected. Spike looks initially rough, crinkled, stunted and finally distorted. In severe cases, spikes become rotten and brittle. The number of flowers in a spike is significantly reduced. During rainy season, the infection is more severe, generally, from June-September and total loss of the second year crop of the 'Calcutta single' cultivar of tuberose is common. Infected bulbs harbour nematode in the scaly leaves outside the bulbs. Nematodes can also survive in the dry scaly leaves, stem and flowers for more than 25 months.

Control:

1. Pre-soaking of bulbs in plain water or in neem seed kernel extract (SSKE, 4%) for overnight followed by dipping in monocrotophos-36SL at 5000 ppm for 4-6 hours is effective.
2. Three to four sprayings of monocrotophos-36SL at 5000 ppm at 15-20 days interval is required after sprouting.
3. Field sanitation is a must for reducing the infection.
4. Infected plant parts should be either burnt or buried into the soil.

Root Knot Nematode

The causal organism is *Meloidogyne incognita*. The infected plants show symptoms of dwarfing, yellowing and withering. There are many galls on the diseased roots. The severely infected plants have thin floral axis and fewer flowers, while the most severely infected plants fail to survive.

Control:

1. Pre-soaking of bulbs in neem seed kernel extract (SSKE 4%) for overnight followed by dipping in monocrotophos-36SL at 5000 ppm for 4-6 hours is effective.
2. Field sanitation is a must for reducing the infection.
3. Infected plant parts should be either burnt or buried into the soil.

Thrips

Thrips which damage and cause distortion of the spikes are the major pests. Thrips suck sap from the leaves and flowers, ultimately damaging the whole plant. Sometimes they are associated with a contagious disease known as bunchy top where the inflorescence is malformed.

Control:

1. Thrips can be managed by spraying the plants with 0.1% Rogor twice at 10 days interval.
2. Dimethoate or oxydementonmethyl 0.05% or Imidaclofrid 0.01% or Cartap hydrochloride 0.075% spray at 15 days interval controls thrips.
3. Malathion and carbaryl are also very effective against these pests.

Aphids

The disease is caused by *Aphid spiraecola*. It is a small green coloured aphid, which sucks the saps of the plant. The aphid attacks the bud and growing spikes. Infected spike is damaged due to fall of buds and flowers from the spike. The effected spikes wither and droop.

Control:

1. Spraying of pongamia or neem oil at early infection level prevents pest multiplication.
2. Spray of dimethoat or oxydemaenton methyl 0.05% or imidaclo-prid 0.01% or acephate 0.075% at 15 days interval controls aphids.

Weevil

The pest is known as *Myllocerus* spp. Adults are ash coloured. They feed on leaves during night hours and are difficult to locate. The symptoms appear on leaves as cuts from margin side.

Control:

1. Fortnightly spray of Malathion 0.1% or Carbaryl 0.2% during evening protects the foliage from weevil damage.
2. Drenching beds with chloropyriphos or Malathion 0.1% reduce grub populations.

Penta Tomid Bug

The pest, *Nezara viridula,* causes damage to flower production. In first stage, eggs are laid in groups on leaf surface. Nymphs are green coloured. Adults suck the sap from basal portion of growing buds and flowers. In severe infection, an affected bud withers and drops off, and causes great loss in flower production.

Control:

1. Hand picking of adults as far as possible and destruction of egg masses checks the damage.
2. Spraying Malathion 0.1% or endosulphon 0.07% controls the bug.

Red Spider Mites

First symptom of spider mite attack on tuberose appears on foliage in the form of yellow stripes and streaks. Spider sucks sap from the foliage. The leaf becomes yellow, silvery and in later stage bronze to cause distortion of foliage.

Control:

Spraying with Kelthane (2ml per litre) has been found very effective to control the mites.

TUBER COLLECTION AND STORAGE

Harvesting of tuberose bulbs at the proper stage of maturity is important. The bulbs reach maturity when the flowering is over and the plant growth ceases. At this stage, the plant growth ceases, the old leaves and bulbs become almost dry. About 100 quintal of bulbs can be harvested from one hectare of field-grown commercial crop. The bulbs should be dug up in November, stored over the winter in airy, dry places. In the spring, they should be separated and replanted in April.

SUMMARY

The tuberose (*Polyanthes tuberose* Linn. Rajanigandha) is a bulbous plant belonging to the family Amaryllidaceae. They are commercially cultivated for cut and loose flower trade, and also for the extraction of its highly valued natural flower oil. Tuberose flower can be successfully grown under wide environmental conditions ranging from tropical to subtropical and temperate climate. Tuberose, a very popular cut flower crop in India, is grown mainly in the eastern part of the country i.e. West Bengal, and also in northern plains and some parts of south. Both single and double flower varieties are popular. It is a popular cut flower, not only for use in arrangements, but also for individual florets that can provide fragrance to bouquets and boutonnieres. It has great economic potential in the cut flower market and in essence-oil industry. It grows in elongated spikes that produce clusters of white flowers that are 5-6 cm long and have a characteristic bending. The flower spikes grow up to 60 cm in length. Tuberose flowers bloom in basipital succession (from the bottom toward the top of the spike). The flowers remain fresh for a long time and are suitable for long distance transportation. Flower yield varies with variety, plant density and planted bulb size and crop, management. The flowers of this plant blossom in the night. The sweet smell lasts till the morning. It is a unique flower - possibly the one which is used in every spheres of the Indian-Hindu life. Important cultivars are Calcutta Single, Calcutta Double, Hyderabad single, Hyderabad double, Phule Rajni, Mexican Single, Sikkim selection, Prajwal, Suvashini, Vaibhav, Arka nirantara and Shringar, Swarn Rekha, Rajat Rekha, which are commercially grown in Assam, Maharashtra, Gujarat, Haryana, Karnataka, Andhra Pradesh, Tamil Nadu, Uttar Pradesh, Uttarakhand, West Bengal and Orrisa.

SUGGESTE READING

1. Banerji BK, Dwivedi AK 2009 Diseases of tuberose and its management, *Floriculture and Nursery Times*, **3** 16-18

2. Tewari A, Banerji BK 2010 Tuberose at a glance, *Floriculture and Nursery Times*, **3** 20-23.

GARDENING TECHNIQUES, TOOLS AND DIRECTORY OF SERVICES

Sanjay Kumar

INTRODUCTION

To set up and maintain a home garden, irrespective of the kind of ornamental species planted, special type of gardening tools and equipments are necessary. This chapter aims to provide basic information regarding gardening tools and equipments, and their maintenance. A list of useful addresses is provided from where these garden tools can be procured.

WATERING EQUIPMENTS

Hose

A hose, made of rubber or plastic, can be stretched according to the requirement. Rubber hoses are durable but are heavy. Now, reinforced plastic hose having single or double beaded linings are available in the market. They are light in weight and do not crack quickly.

Watering Can

Watering cans, made of plastic with 5 to 10 litre capacity, are available in the market. They are light in weight and easy to handle. Each contains a removable nozzle with a fine rose, which can be adjusted according to the requirement of water.

Sprinkler

Two types of sprinklers are available; rotating and oscillating type. Oscillating type of sprinklers cover a larger area than rotating type and can be adjusted to cover the desired area by controlling the water pressure.

Sprayers

Sprayers, used for applying insecticides and fungicides, are available in different capacities and sizes. It ranges from small hand held sprayer to large power sprayer capable of delivering 50 or more litres of spray per minute. The small hand held sprayer made of plastic or metal body with varying capacity can spray 100 ml to 2 litres at a time. Similarly, electric or petrol operated sprayer made of stainless steel piston can deliver 12-100 litres of liquid per minute and are suitable for greenhouse, orchards and plantation crops.

Knapsack Sprayer

Its capacity ranges from 10-16 litre and can be carried on the operators back. Pressure is maintained by pressing a lever up and down with one hand while spraying is done with the other hand. Its body is made of high density polyethylene, which is light in weight and comfortable for the operator. In knapsack type, petrol operated power sprayers are also available. The capacity of the tank is around 20 litres. Although power sprayer is expensive in comparison to hand-operated sprayer but it can cover very large area in a very short time.

Foot Sprayer

It is ideal for all types of spraying and is suitable for ornamental gardens, orchards and plantation crops. Its body is made of brass, attached with 8 m long delivery hose along with spray gun or lance. It can be operated very easily.

SOIL WORKING TOOLS

All garden lovers need five basic soil working tools i.e, spade, fork, rake, hoes and khurpi.

Spade

Spades are most commonly used garden tools. Front of the spade is wide and sharp, and back is round and heavy on which foot can be placed to exert pressure. It is used mainly for breaking up hard or compact soil during bed preparation, making irrigation channels and installation of all types of plants.

Pick

It is pointed at one end and flat at the other end. Its pointed end is used for digging and breaking up hard rocky soil and flat end is used for removing stones from the soil.

Fork

In this garden tool, 4-5 long pointed iron teeth are fixed at one end and a round top at other end for fixing iron or wooden foot. It is generally used for breaking clods in soil and preparing planting beds that are not too deep.

Rake

A rake is made of iron with strong widely spaced teeth. It is used for smoothening the soil surface for preparation of seed bed. It is also used for gathering weeds or leaves and removing grass from lawn.

Hoe

There are long-handled and short-handled hoes. The short handled hoe is frequently used in vegetable gardening for seed sowing. It is also used for weeding.

Khurpi

It is the most frequently used gardening tools. Its front edge is 2-3 inch wide and sharp and back end is round with wooden or iron grip. It helps in weeding and earthing of plants.

PRUNING TOOLS

All the plants, trees, shrubs or climbers need pruning or cutting of some parts during their growth. It is necessary to remove unwanted and diseased branches. There are different types of tools available for pruning such as pruning knife, secateurs, lopping shears, hedge shears, pruning saw and power operated hedge trimmer and saw.

OTHER GARDEN AIDS

Various other garden tools like budding and grafting knife, hand operated lawn mower, power operated lawn mower, brush cutter and wheel barrow help gardeners in a great way to maintain their lawn and garden plants.

CARE AND MAINTENANCE OF EQUIPMENT

It is necessary to maintain the garden tools and equipments in good shape for smooth operation and efficient working. Faulty equipments damage the shape of plants and visual quality of lawn grass, and also are unable to effectively control plant diseases and pests. Whether one is using a small sprayer or a power-operated equipment, regular cleaning to remove dust and dirt, and proper lubrication is recommended to prevent them from sticking and squeaking. While using sprayers, use clean water, free from dirt and other materials, to avoid cloging the nozzles so that high pressure does not cause sudden spillage of insecticides, which may harm the operator.

Gardening tools and accessories

Gardening tools

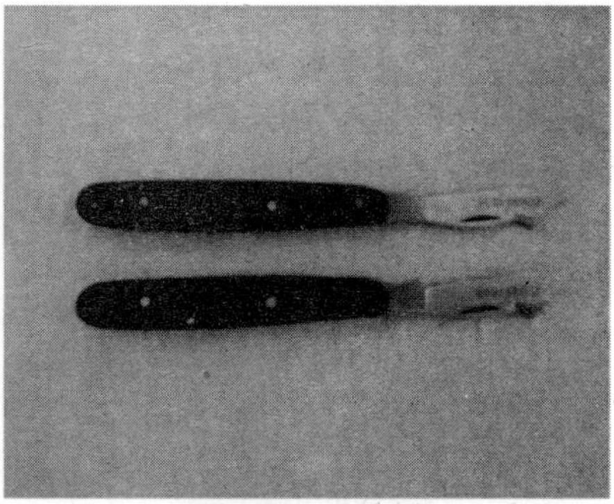

Budding knife

Garden sprayers

Fig. 26. Gardening tools

LIST OF RETAILERS IN HIMACHAL PRADESH

1. Agriculture House
Near B.D.O Office,
Nagrota Bagwan,
Kangra-176047
9816162650

2. Neckraj Agro
Yol Cantt, Dharamsala,
Kangra-176052
9816781112

3. Miar Enterprises
Sheela Chowk, Dharamsala
Kangra-176057
01892-225124/9418427115

4. Walia Agro Sale centre
Main Bazar, Palampur
Kangra-176061
01894-232859

5. Krishi Sewa Kendra
Near Sainik Rest House,
Palampur, Kangra-176061
01894-235878/9418991616

6. Basudev Krishak Seva Kendra
HRTC Complex, New Bus Stand,
Palampur
Kangra-176061
9218611123/9418079263

7. Avishanka Intertraders Pvt Ltd.
Near New Bus stand, Palampur
Kangra-176061
01894-230193/9816347110

8. General Sales Agency
Near New Bus stand, Palampur
Kangra-176061
01894-230173/9816047047

9. Jain & Jain
Old Bus Stand, Palampur
Kangra-176061
01894-32528/9816032528

10. Jain Irrigation Pvt. Ltd.
Kalara House, Near Yamini
Hotel, Palampur
Kangra-176061
01894-23 4152/94180 12733

11. Raghu Agro Agencies
Khaira Road Chowk, Bhawarna,
Palampur, Kangra-176083
01894-247434/9816048434

12. Shiv Shakti Traders
Maranda, Palampur
Kangra-176102
01894-248446/9875213823

13. Ladohia Agro Sales
Near Block Gate, 2nd Floor,
Panchrukhi, Palampur
Kangra-176103
9418122877

14. Kisan Beej Bhandar
Ayurvedic College Road,
Paprola, Baijnath
Kangra-176115
9418248703

15. Sharma Beej Bhandar
Main Bazar Baijnath
Kangra-176125
9418105012

16. Jaryal Agencies
Chogan Bazar, Nurpur
Kangra-176202
01893-283820/9816460747

17. Rishu Seed Store
Main Bazar, Rait
Kangra-176208
9418165520

18. Interstate Agro Marketing
System, Ramayan Kunj
Below Auckland Hotel
Shimla-171001
01792-221468

19. Kamal Komal & Co.
62, The Mall
Shimla-171001
9418072285

20. Jain Irrigation Pvt. Ltd.
C-69, Second Floor,
Phase-1, Sector-1
New Shimla -171001
0177-267 2513/94181 71333

21. Adani Agri Fresh Ltd.
1-New Road
Chhota Shimla
Shimla-1710012
9816675333

22. Rohini Enterprises
Hotel Broadway, Sanjauli
Shimla-171006
0177-2842938/9418344721

23. Shimla Agro Equipments (P)
Ltd.
Ground Floor,
Hotel Navrattan, Dhalli
Shimla-171012
9418033095

24. Ajay Enterprises
Bus Stand, Theog
Shimla -171201
9816638096

25. Shimla Machinery Stores
Bus Stand,Theog
Shimla-171201
9418055455

26. Mahindra & Co.
Main Bazar,Theog
Shimla-171201
9459343471

27. Tajta Enterprises
Main Bazar, Kotkhai
Shimla-171202
9816055770

28. Lok Sewa Agro Services
Opp. Bus Stand, Rohru
Shimla-171207
9816025933/9418040354

29. Kehrumal Agyaram
Main Bazar, Rohru
Shimla-171207
9418040120

30. Shiva Agrochemicals
Near Bus Stand, Rohru
Shimla-171207
9418452295

31. United Traders
Bus Stand, Rohru
Shimla-171207
9418040354

32. Lok Sewa Agro Services
Opp.Bus Stand, Rohru
Shimla-171207
9418040354

33. Rawat Agro Services
Main Bazar, Chargaon
Shimla-171208
9816638096

34. Narkanda Sales Corporation
Shopping Complex, Bus Stand
Narkanda, Shimla-171213
01782-242408/9816042408

35. Krishan Dev Sharma & Bros.
Bus Stand, Narkanda
Shimla-171213
01792-224710/9418230617

36. Sharma Seeds & Insecticides
NH-22, Verma Coplex, Shoghi
Shimla-1712139
9218573120

37. Deepak Agro Chemicals
Near Bus stand, Rampur Bushar
Shimla-172001
01782-233217/9218702592

38. Green Effect Farm
Near Bus stand, Rampur Bushar
Shimla-172001
9418032306

39. Mehta Enterprises
Jawalda, Nankhari
Shimla-172021
9418051801

40. Jai Mata Traders
Near Canara Bank,
Opp. Bypass, Mangal Market,
Baddi, Solan-173212
01792-221861/9816075144

41. Suri Enterprises
Main Bazar, Mall Road
Solan-173212
9816904887/9816167235

42. C.M. Seeds Farm
Lower Market, Near old Bus
stand, Solan-173212
9882162408

43. Kavita Seeds & Pesticides,
26, Murari Market, The Mall
Solan-173212
9816077468

44. Agroline India
Shop No. 35, 2nd Floor,
The Mall, Solan-173212
9418055990

45. Greenfield Agri Solutions
Near Tel. Exchange,
Jawahar Park, Solan-173212
9816088109/9418048535

46. Ashu Sharma
Palace Road, Solan-173212
9857065550

47. Jain Irrigation Pvt. Ltd.
H.No. 146/3, Ward no. 11,
Surya vihar, Rajgarh Road,
Solan-173212
94591 35912

48. Shivalik Agro Engineers
Near DAV School
By Pass Road, Solan-173213
01792-221861/09816075144

49. Alisha Enterprises
Mall Road, Kandaghat
Solan-173215
9816080580/9805080579/581

50. Shivalik Agro Chemicals
Mall Road, NH-22, Kandaghat
Solan-173215
9816362914

51. Vishal Motors & Engineers
Old Bus Stand
Mandi-175001
9418082431

52. Durga Seeds Farm
Ner Chowk, Mandi-175008
9418008677

53. GreenTech Agri. Sector Pvt. Ltd.
Near Escort Agency, Ner Chowk
Mandi-175008
9817096104

54. Saini Electricals
Ner Chowk
Mandi-175008
9816003114/9817606014

55. Kamal Seed House
Bus Stand, Karsog
Mandi-175011
9816163489

56. Aprajita Enterprises
Main Bazar, Churag, Karsog
Mandi-175011
9816167235

57. K.D Seeds House
Shop No. 10,Indira Market
Sundernagar, Mandi-175019
9418065375

58. Jain Irrigation Pvt. Ltd.
286/10, BBMB Colony Road
Sundernagar
Mandi-175019
01907-265 621/94181 69333

59. Om Seed House
Shop No.-199,Indira Market
Sundernagar, Mandi-175019
9418078304

60. Saini Agro Services
Near Milk Plant,Chakkar
P.O. Gutkar
Mandi-175021
9816362914

61. Harvel Agua India Pvt. Ltd.
9th KM Stone,Vill.-
Bagla,N.H.21
P.O.- Nagchala
Mandi-175021
01905-241986/9816008973

62. Awasthi Enterprises
Chauntra, Jogindernagar
Mandi-175032
9816163489

63. Raghav Enterprises
Sarwari bazaar
Kullu-175101
9816022324

64. Vyas Seed
Bhunter
Kullu-175101
9418065375/

65. Agro Sales Corporation
New Akhara Bazar
Kullu-175101
01902-265375/9882208484

66. Mahant Enterprises
Akhara Bazar
Kullu-175101
9882034735

67. Orchards Enterprises
Ram Bagh, Akhara Bazar
Kullu-175101
01902-222666/9816032222

68. Narayan Agro Sales Centre
Main Bazar, Jari
Kullu-175105
9805110149

69. Choudhary Seeds & Pesticides
Near State Bank Of India,
Bhunter
Kullu-175125
9816066275

70. Sangam Agro Service
Near State Bank Of India,
Bhunter
Kullu-175125
01902-267142

71. Radha Agro Chemicals
Main Bazar, Bhunter
Kullu-175125
01902-265112

72. Arvind Agro Sales Service
Ropa Road, Bajaura
Kullu-175125
9816290330

73. Jain Irrigation Pvt. Ltd.
Gaon Piradi, Opp.Angora
Breeding Farm
Mohal, Kullu-175126
1902-261311/ 9418073923

74. Lucky Pesticides
Vill. & P O Raison, Raison
Kullu-175128
9816187590

75. Sharma Kisan Seva Kendra
Vill. & P O Raison
Kullu-175128
9805730377

76. Nirmal Agro Enterprises
PatliKuhl, Manali Road
Kullu-175128
9418008677

77. Siddarth Agro sale
Patli Kuhl, Katrain
Kullu-175129
9418105399

78. Chopra Enterprises
Naggar Road, Patli Kuhl
Kullu-175129
01902-241045

79. Hira Traders
Naggar Road, Patli Kuhl
Kullu-175129
01902-240045/240467

80. Thakur Seeds & Pestcides
Dwara, Dobhi, Kullu-175129
01902-240584

81. Balaji Agri Clinic
Shop No.13, Market Complex
Near Bus Stand,
Bilaspur-174001
9817096104

82. Global Agritech
Shop No. 2G, Upper Market
Bilaspur-174001
9816003114/9817606014

83. Jain Irrigation Pvt. Ltd.
House no. - 96/13,
Roura Sector - III
Bilaspur-174001
94181 91511

84. Crop Solutions
Dadhol, Ghumarwin
Bilaspur-174023
9816073229

85. Farmers Fertilizers
P.O Berthin, Jhandutta
Bilaspur-174029
9418078304

86. Saini Seed House
Main Bazar Jahu, Bhoranj
Hamirpur-176048
9816163489

87. Jain Irrigation Pvt. Ltd.
Ward no. 8, H.No. 117,
Nayanagar, Hamirpur-177001
9418025833

88. Kishan Beej Bhandar
Near New Bus stand, Nadaun
Hamirpur-177033
9418082267

89. Gupta Agro Sales & Service
Centre
Majra, Ponata Sahib
Sirmour-173021
9318827260

90. Panwar Trading Company
Badril Nagar, Kishanpura Road
Ponata Sahib
Sirmour-173025
9418082431

91. Jain Irrigation Pvt. Ltd.
Ward no.6, Bangran Road.
Ponata Sahib
Sirmour-173025
9418013833

92. Lovely Seeds
Gagret Road, Kala Amb
Sirmour-173030
9816060804

93. Vijay Enterprises
Near UCO Bank, Rajgarh
Sirmour-173101
9816163489

94. K.K. Agri Hitech services
Near Giripul Barrier,
Yashwant Nagar, Rajgarh
Sirmour-173223
9816845101

95. Neegam Singh & Bros.
Main Bazar, Reckong Peo
Kinnaur-172107
9418046068

96. Janni Plant Health Centre
Main Bazar, Reckong Peo
Kinnaur-172107
8894580080

97. Sunninder Kumar
Opp.New Sada Complex,
Main Bazar, Reckong Peo
Kinnaur-172107
9418358469

98. Saini Seed Store
Near SBI, Main Bazar
Una-174303
9418178227

99. Agriculture General Store
Bachat Bhawan, Jassal Market
Una-174303
9418338538

100. Jain Irrigation Pvt. Ltd.
Jaswal Complex, 1st floor
Opp. DC Office
Una-174303
94188 89640

101. Himalaya Irrigation System
Lal Singhi, Raisarai
Una-174303
9816683877

102. Hindustan Agro Chemicals
Main Bazar, Punjabo
Una-177208
9816372129

103. Kumar & Company
Main Bazar, Galuwal Chowk
Una-177209
9816322028

104. H.P. Farmers
1st Floor, Dogra Bazar
Chamba-176310
9816683097

105. Jain Irrigation Pvt. Ltd.
Mohalla Hardaspura
Near Graveyard
Chamba-176310
9459046807/ 9459135908

LIST OF RETAILERS IN UTTRAKHAND

1. Agarwal Auto Traders
Ramnagar Road, Kashipur,
Udham Singh Nagar-244713
05947- 270056/274216/
98976 51232

2. Kanhaiyalal Babu Ram
Near Roadways Opposite Deepak
Cinema, Kashipur, Udham
Singh Nagar-244713
05947-275056

3. Farm Friends
41, Pant Market, Kashipur,
Udham Singh Nagar-244713
05947- 274758/ 9837048256

4. Kissan Fertilizer Agency
Main Road, Opp- Chiti More,
Kashipur,
Udham Singh Nagar-244713
05949-281292/ 9837031292

5. Shanti Prasad & Sons
Main Chouraha Bazpur Road,
Kashipur ,
Udham Singh Nagar-244713
9837891691

6. Astha Beej Pvt. Ltd.
B-2, Industrial Estate,
Bazpur Road, Kashipur ,
Udham Singh Nagar-244713
05947-262047/9412089997

7. B. P. Agrochemicals
B-2, Industrial Estate,
Bazpur Road, Kashipur;
Udham Singh Nagar-244713
05947-274065/ 275467/
9837095965

8. Continental Seeds &
Chemicals Ltd.
Ram Raj Road, Bazpur,
Kashipur,
Udham Singh Nagar-244713
05947-281227/
9810071972

9. Quality Tools
Civil Lines, Roorkee,
Haridwar-247667
94113 71750

10. S.Kumar & Company
15, Nahar Kinara, Roorkee,
Haridwar-247667
01331-247667/ 9837076949

11. Shree Balaji Sales Corporation,
Shop.No-9, Super Market,
Rampur Tiraha Chungi,
Roorkee, Haridwar-247667
9412231084

12. Nidhi Enterprises
36, Arhat Bazar,
Dehradun-248001
9639000070/9412972132

13. Agarwal Fertilizer
Opp. Subzi Mandi, Sharanpur
Chowk, Dehradun-248001
9319818923

14. Harvel Agua India Pvt. Ltd.
4/13, Rajpur road, Behind Moti
Mahal, Dehradun-248001
089799 66600

15. Jain Irrigation Pvt. Ltd.
H-333, Nehru Colony, Near
Police Station, Dehradun-248001
94129 97917

16. Advance Engineering Co.
175/B-2, Rajpur Road,
Dehradun-248009
0135-2733783/9412050842

17. Chopra International
Railway Road, Opp. Bank of
India, Dehradun-248001
0135-2648485

18. Indo Agrotech
No.75/1, Dhamawala Bazar,
Dehradhun-248001
0135-2624556

19. Ahinsa Agriculturals Pvt. Ltd.
SRE, Kalsia Road, Saharanpur,
Dehradhun-248001
0135-2654920

20. Indo German Scientific
Research Institute, No.19,
Uttaranchal Complex, Haridwar
Road, Dehradhun-248001
01352666215 9837067124

21. Vishnu Khad Bhandar
Khatima Road, Sitarganj,
Udham Singh Nagar-262308
05948- 254198/ 9412038490

22. Nav Durga Agro Chemical
Nanital Road, Bazpur, Udham
Singh Nagar-262401
05949-288158/9412089338

23. Singhal Automobiles
Kashipur Road, Ramnagar,
Nanital -263002
05947-251480/ 9837041538

24. Sanwal Seeds
Haldwani, Nanital -263139
05946-284651/ 9412086543

25. Kichha Agro Agencies
By Pass Chouraha, Bareilly Road, Kichha,
Udham Singh Nagar-263148
05944-264621/264031/ 9837790619

26. R.S.Fertilizers
Bareilly Road,Kichha,
Udham Singh Nagar-263148
05944-265511/ 9927066511

27. Satyapal Rajeev Kumar
Gadarpur,
Udham Singh Nagar-263152
05949-271147/ 9719700047

28. Aajanta Fertilisers
Galla Mandi Road, Rudrapur,
Udham Singh Nagar- 263153
05944-243216/ 241837/ 9927173994

29. Uttranchal Agrisystems
Netaji Subhas Marg, Rudrapur,
Udham Singh Nagar- 263153
9719015500/9719022500

30. Munjal Seeds & Pesticides
B-16, Gole Market,
Opposite Gurudwara, Rudrapur,
Udham Singh Nagar- 263153
9837667858

31. Layalpur Traders
Galla Mandi Gate,
Udham Singh Nagar- 263153
05944-243505/ 9412088014

32. Wimco Seedlings Ltd.
Vill. Bagwala,
Kashipur Road, Rudrapur,
Udham Singh Nagar- 263153
05944-261773/ 261760

33. T.Stanes
Compound of Rungta Industries,
Kashipur road, Rudrapur,
Udham Singh Nagar- 263153
05944-247142

LIST OF RETAILERS IN JAMMU & KASHMIR

1. R.B. Seeds Farm
Shop No. 10, JDA Complex
New Sabzi Mandi, Narwal,
Jammu-180 001
9419182480

2. Navartis India Ltd.
5, Trikuta Nagar,
Jammu-180001
0194-2436211

3. Quality Pesticides
Kanak Mandi,
Jammu-180001
0194-2541448

4. Shivani Pest Control Service
Guoshalla Complex, Amphalla,
Jammu-180001
0194-2573091

5. Mahajan Seed House
Exhibition Ground,
Near Pest Club,
Jammu-180001
0191-2573963/9419224334

6. Saraswati Agro Chemicals (India) Pvt. Ltd.
Lane 2, Phase 1, Sidco Indl. Complex, Bari Brahmna,
Jammu-181133
01923-221433/9906903065

7. Harvel Agua India Pvt. Ltd.
364, Sector 7 Road,
Channi Himmat,
Jammu-180015
9596730396

8. Atwal Abrol Agro
Main Bazar, Ramnagar
Udhampur-182122
9419214507

9. Geeco Sales Corp
Karan Nagar, Srinagar-190001
0191-2470191

10. Narsingh Agro Agency
Main Bazar, Khagwal,
Samba-184121
9419151388

11. Gulshan & Co
Bala Garden,
Srinagar-190001
0191-2470515

12. Khan Agri Implements Industries
Nati Pora Chowk,
Srinagar-190001
0191-2433278

13. Kashmir Pesticides
Pahalgam Road,
Srinagar-190001
0191-234438

14. Fungicide India Ltd.
Sheikh Bagh,
Srinagar-190001
0191-2450999

15. Diamond Pesticides
Bhat Building,
Near Super Bazar,
Srinagar-190001
0191-2458795

16. Hakim & Co.
Court Road,
Srinagar-190001
9419079519

17. Shafaat Naqshbandi & Co.
Main Road,
Srinagar-190001
9419019797

18. Hanief Motors
Maulana Azad Road
Srinagar-190001
0194-2454600/9419011444

19. JK Krishi Vikas Cooperative Ltd.
Lal Mandi, Paraypora,
Srinagar-190008
94190 02817

20. Premier Sales Agencies
43 SMC Building H S High Street
Srinagar-190009
9419063924

21. Ashique Agro
Main Bazar,
Anantnag-192101
9797960235

22. N. Bhat Traders
Pahalgam Road,
Zirpora, Bijbehara,
Anantnag-192124
9906721627

23. Unique Enterprises
Main Bazar, Pulwama-192301
9419570058

24. AL Noor Enterprises
Court Road,
Pulwama-192301
9419031592

25. Kashmir Agro
Main Bazar,
Shopian-192303
9596381327

26. Zafar Traders
Near Bus Stand,
Baramulla-193101
9419556829

27. Endofil House
Main Bazar,
Baramulla-193101
9419305181

28. M.A Enterprises
Main Bazar,
Baramulla-193101
9419864403/9419041123

29. Greenland Products
Main Bazar, Rahma Block,
Baramulla -193101
9419577209

30. Crown Enterprises
Main Bazar, Dingiwacha Block,
Baramulla-193101
9419716660

31. Khuroo Pesticides
Tehsil Road,
Sopor-193201
9419038040

32. Q.F. Enterprises
Mohalla Lala Baba Sahib
Sopor -193201
9906683483

33. Mir Enterprises
Marama, Sopor-193201
9906755649

34. Wani Traders
Main Bazar, Handwara Block,
Kupwara-193221
9697973700

35. B.R. Agrotech Ltd.
Sidco Industrial Area
Kathua-184101
01922-203025/9366525817

36. Bharat Insecticides Ltd.
Sido Industrial Area
Kathua-184101
01922-232165

37. Antriksh Pesticide Store
Ward No. 6,
Opposite Moon light Cinema,
Kathua-184 101

38. Kisan Agro
Dialachak
Kathua-184101

39. Verma Agro Chemicals
Nagri, Kathua-184 101

40. Krishna Seed & Pesticides Store
Village-Dhani Belra
Kathua-184 101

41. Raghu Pesticides
New Sabzi mandi,
Shop No. 14 & 15
Kathua-184 101

42. Khajuria Hardware
Chadwal More
Kathua-184101

43. New Kisan Seed Store
New Foot Bridge,
Kathua-184101

44. Mehta Pesticides
Phalote, Kathua-184101

45. Pragati Pesticides
Kalibari, Janglote,
Kathua-184104

46. Baba Pesticides
Haripur More, Hiranagar,
Kathua-184142

47. Krishi Vikas Kendra
Kootah, Hiranagar,
Kathua-184142

48. Krishan Lal
Sanji More, Rasila Market,
Hiranagar,
Kathua-184142

49. Raju Fertilizer & Pesticide
Sanji More, Hiranagar,
Kathua-184142

50. Shub Kumar Vaid
Mahanpur, Basoli,
Kathua-184201

51. Mahajan Pesticides
Mandli, Billawar,
Kathua-184202

52. Vikas Basotra
Village -Phinter,
Bhaddu, Billawar,
Kathua-184203

53. Naresh Kumar
Kohag, Billawar,
Kathua-184203

54. Jain Irrigation Pvt. Ltd.
Near Income Tax Office
Palampur, Kangra - 176 061
1894-234152/9418016033